Learning Materials in Biosciences

Learning Materials in Biosciences textbooks compactly and concisely discuss a specific biological, biomedical, biochemical, bioengineering or cell biologic topic. The textbooks in this series are based on lectures for upper-level undergraduates, master's and graduate students, presented and written by authoritative figures in the field at leading universities around the globe.

The titles are organized to guide the reader to a deeper understanding of the concepts covered.

Each textbook provides readers with fundamental insights into the subject and prepares them to independently pursue further thinking and research on the topic. Colored figures, step-by-step protocols and take-home messages offer an accessible approach to learning and understanding.

In addition to being designed to benefit students, Learning Materials textbooks represent a valuable tool for lecturers and teachers, helping them to prepare their own respective coursework.

More information about this series at http://www.springer.com/series/15430

Cornelia Kasper • Dominik Egger •
Antonina Lavrentieva
Editors

Basic Concepts on 3D Cell Culture

 Springer

Editors
Cornelia Kasper
Institute of Cell and Tissue Culture
Technologies
University of Natural Resources and
Life Sciences
Vienna, Austria

Dominik Egger
Institute of Cell and Tissue Culture
Technologies
University of Natural Resources and
Life Sciences
Vienna, Austria

Antonina Lavrentieva
Institute of Technical Chemistry
Leibniz University of Hannover
Hannover, Germany

ISSN 2509-6125   ISSN 2509-6133   (electronic)
Learning Materials in Biosciences
ISBN 978-3-030-66748-1   ISBN 978-3-030-66749-8   (eBook)
https://doi.org/10.1007/978-3-030-66749-8

This Springer imprint is published by the registered company Springer Nature Switzerland AG.
The registered company address is: Gewerbestrasse 11, 6330 Cham, Switzerland

Preface

Cell culture technology is a continuously developing field in biotechnology as well as in basic sciences. Apart from the isolation of new cell types and the development of new cell lines, there is tremendous progress in cultivation platforms, media optimization, novel arrays and monitoring techniques as well as increasing complexity of *in vitro* models. Especially, three-dimensional (3D) cell cultivation becomes a leading edge of this development. 3D cell cultivation develops in various directions with growing complexity, including co-cultivation of different cell types in 3D, spatial distribution, and precise geometry control by the implementation of bioprinting, as well as dynamic cultivation of 3D cellular constructs in bioreactors and microfluidic systems. If in our previous book ("Cell Culture Technology" Springer Learning Materials in Biosciences. Eds: Prof. Dr. Cornelia Kasper, Dr. Verena Charwat, Dr. Antonina Lavrentieva, Springer Verlag, 2018) we concentrated on modern, but mainly classical, 2D cell cultivation techniques, this volume presents a logical continuation and extension of cell culture in the third dimension. We included in this book various aspects and techniques of 3D cell cultures. Being more physiologically relevant, 3D *in vitro* systems become increasingly challenging in terms of monitoring, viability evaluation, and choice of suitable cultivation platform. For this book, we brought together leading experts in various fields of 3D cell cultures: biotechnologists, biologists, material scientists, and engineers, who shared their valuable knowledge and described the current state of the art in their fields of expertise.

In Chap. 1 of this book, a general and profound overview and comparison are given over the main aspects for "translating" the cell culture techniques from traditional 2D to advanced 3D cell culture conditions. Special focus is laid on aspects important within the field of translational research and highlighting the essential criteria and components and stimuli involved in shaping the cellular microenvironment. Furthermore, the different methodologies for implementing 3D cultures are described followed by a demonstration of various applications.

Nowadays, a wide spectrum of different products and equipment exists for cell culture applications. Chapter 2 presents the "essentials" for 3D culture application products.

Numerous examples are presented along with a special focus on scaffolding techniques as well as on different bioreactors systems for dynamic cultivation.

Chapter 3 presents the perception of 2D and 3D cell culture conditions from the cell perspective. On the example of isolation and cultivation of mesenchymal stem/stromal cells, protocol steps are discussed: what is the natural *in vivo* cell microenvironment, what happens to the cells during isolation step and following *in vitro* cultivation, and how 3D cell cultures and co-cultures bring the isolated cells into physiologically "comfort" *in vitro* environment. This chapter also includes a commix, which demonstrates in entertaining way all discussed issues.

Biologists, chemists, and material scientists have developed a wide range of 3D scaffolds to allow 3D *in vivo*-like extracellular environment. Chapter 4 of our book summarizes the scaffold material design techniques, for both synthetic and natural scaffolds. Moreover, the application of natural decellularized soft and hard tissues is described in this chapter. The authors subdivide scaffold materials, those which are used for cell entrapment versus those where porous matrix is manufactured first with subsequent population by the cells.

Hydrogels, 3D crosslinked networks of polymers, have become a very popular 3D cultivation platform since they provide researchers with a versatile tunable toolbox allowing the creation of 3D constructs with a great variety of optical, mechanical, chemical, and biological properties. Chapter 5 describes hydrogel classifications, mechanical and physical requirements of hydrogels, as well as methods of hydrogel characterization, sample recovery, and cell analysis. Furthermore, the creation of gradient hydrogel constructs and the field of hydrogel applications such as microfluidics and bioprinting are described in this chapter.

To reconstruct the microvasculature in 3D *in vitro* models and tissue-engineered constructs, various techniques were developed over the past years. On the one hand, increasing sizes of tissue-engineered constructs require an adequate supply of nutrients and oxygen. On the other hand, 3D vascularized aggregates represent physiologically relevant *in vitro* models for drug screening and disease models. Chapter 6 summarizes procedures and aspects for the isolation and characterization of vascular cells and the generation of various 3D vascular structures. The authors present detailed protocols for the creation of functional vascularized organoids.

Over the last 10 years, a rapidly growing number of approaches were developed for scaffold-free 3D culture. Many research groups but also companies worked on the establishment of more or less self-zing concepts. Nowadays, spheroid cultures are gaining increased interest and are applied much more widely, especially due to a growing number of smart materials being available to support the easy handling of these advanced cell cultures. Chapter 7 summarizes a selection of definitions, techniques, and products for realizing scaffold-free 3D cell culture construct cultivation.

In Chap. 8, experts in microfluidics describe how microfluidic techniques can be used to further enhance the physiological relevance of 3D cell cultures. Indeed, the implementation of microfluidic devices allows not only dynamic cultivation of 3D cell cultures but it also gives an opportunity to apply time-resolved concentration profiles of, e.g. growth factors. Another major advantage of microfluidics is the automatization of typical lab procedures, such as mixing, separation, amplification, and detection of components. Both, soft lithography and high-resolution 3D printing methods for microfluidic systems fabrication are also discussed in detail in this chapter. In addition, the authors provide a detailed overview on "organ-on-a-chip" systems, an innovative approach which allows the creation of interplay between organs on miniaturized and automatized platform.

To even more advance the options of 3D cultures, bioprinting technology has been introduced. This technique is based on previously developed 3D additive manufacturing, where 3D models are printed layer by layer. With the help of bioprinting, precise spatial cell and biomaterial distribution in 3D constructs can be achieved. Moreover, 3D constructs with complex geometry can be created, allowing the manufacturing of in vivo-like tissues and organs. Chapter 9 of this book gives a comprehensive overview on bioprinting chains, including 3D model generation, natural and synthetic materials used as bioinks, 3D bioprinting methods and bioprinters, as well as maturation and the application of bioprinted tissues.

One of the major limitations in 3D cell cultures is still the analytical techniques for reliable and sensitive "monitoring" of cell and 3D tissue growth. Common to all 3D cell culture applications is the demand for non-destructive methods. Another demand on the "wish list" of researchers and companies involved in product development for cell-based assays and therapies are tools and techniques that can be applied label-free. Most of the microscopic methods are limited with regard to depth and also with demand for transparent materials. In Chap. 10, a profound overview on Raman microscopy and confocal microscopy is presented as examples for systems that are being applied also in 3D cell culture monitoring.

The editors would like to thank all authors contributing to this comprehensive introduction to 3D cell culture technologies and applications textbook. It is out of question that we have all been working under challenging conditions since the beginning of 2020. The COVID-19 crisis has influenced massively not only our work but also our everyday life. The editors are tremendously thankful to the contributing colleagues for dedicating their time to this modern textbook on advanced 3D cell culture.

Hannover, Germany   Antonina Lavrentieva
Vienna, Austria   Cornelia Kasper
October 2020

Abbreviations

ADME	Adsorption, distribution, metabolism, and excretion
ADME-Tox	Adsorption, distribution, metabolism, excretion, and toxicity
hAD-MSCs	Human adipose-derived mesenchymal stem cells
AFM	Atomic force microscopy
AJ	Adherens junctions
AoC	Animal-on-a-chip
APS	Ammonium persulfate
ARCs	Adult rat cardiomyocytes
ASCs	Adult stem cells
ATMPs	Advanced therapy medicinal products
BMP	Bone morphogenetic protein
CAD	Computer-aided design
CAMs	Cell adhesion molecules
CBT	Cell-based therapies
CDM	Contractile differentiation medium
CDMA	Methacrylated cyclodextrin
CHAPS	3-[(3-cholamidopropyl)dimethylammonio]-1-propanesulfonate
CHO	Chinese hamster ovary
CRC	Colorectal cancer
CSFs	Colony stimulating factors
cSMCs	Contractile smooth muscle cells
CT	Computed tomography
CT-RamSES	Cell Tool Raman Statistical Analysis Software
cytoD	Cytochalasin D
DAPI	$4',6$-diamidino-2-phenylindole
DEAE	Diethylaminoethyl
Dex-GMA	Glycidyl methacrylate derivatized dextran
DMSO	Dimethyl sulfoxide
DNA	Deoxyribonucleic acid
DNase	Desoxyribonuclease
DoC	Disease-on-a-chip

DoF	Degree of functionalization
DP	Dermal papilla
DS	Dextran sulfate
DTT	Dithiothreitol
EB	Embryoid body
ECs	Endothelial cells
ECM	Extracellular matrix
ECGS	Endothelial cell growth supplement
E. coli	Escherichia coli
EDTA	Ethylenediaminetetraacetic acid
e.g.	exempli gratia
EGF	Epidermal growth factor
EGTA	Ethylene glycol-bis(β-aminoethyl ether)-N,N,N′,N′-tetraacetic acid
EMA	EU's European Medicines Agency
hEPC	Human endothelial progenitor cell
ESCs	Embryonic stem cells
EtO	Ethylene oxide
FAK	Focal adhesion kinase
FACS	Fluorescence-activated cell sorting
FCS	Fetal calf serum
FDA	U.S. Food and Drug Administration
FFF	Fused filament fabrication
FGF	Fibroblast growth factor
Fig	Figure
FUE	Follicular unit extracts
GAGs	Glycosaminoglycans
GelMA	Gelatin Methacryloyl
GSK-3	Glycogen synthase kinase-3
HA	Hyaluronic acid
HAFs	Human artery-derived fibroblasts
HAMA	Methacrylated hyaluronic acid
HCA	Hierarchical cluster analysis
HepMA	Methacrylated heparin
hESCs	Human embryonic stem cells
HF	Hollow fibers
HHP	High hydrostatic pressure
HHSteC	Human hepatic stellate cells
HIF-1 α	Hypoxia-inducible factor 1 alpha
h(i)PSCs	Human (induced) pluripotent stem cells
HoC	Human-on-a-chip

hpAs	Human primary astrocytes
hpBECs	Human primary brain endothelial cells
hpPs	Human primary pericytes
HRP	Horseradish peroxidase
HSA	Human serum albumin
HUVECs	Human umbilical vein endothelial cells
ICS	Intracapillary space
i.e.	id est
Ifs	Interferons
IGF-I	Insulin-like growth factor
ILs	Interleukins
iPSCs	Induced pluripotent stem cells
IVD	Intervertebral disc cells
LCST	Lower critical solution temperature
LIFT	Laser-induced forward transfer
L-L PS	Liquid–liquid phase separation
LPA	Lysophophatidic acid
LSEC	Liver sinusoidal endothelial cells
MA	Methacrylated
MCs	Mural cells
μCCAs	Micro-cell culture analogs
μTAS	Micro Total Analysis System
MFDS	South Korea's Ministry of Food and Drug Safety
MFU	Minimal functional unit
MI	Myocardial infarction
MOCs	Multi-organ-chips
MRI	Magnetic resonance imaging
MSCs	Mesenchymal stem cells, mesenchymal stromal cells
MWCO	Molecular weight cut-off
NMCs	Neonatal mouse cardiomyocytes
NRCs	Neonatal rat cardiomyocytes
NSCs	Neural stem cells
O_2	Oxygen
OoC	Organ-on-a-Chip
Pa	Pascal
PAA	poly(acrylic acid)
PBMCs	Peripheral blood mononuclear cells
PBS	Phosphate-buffered saline
PBPK	Physiologically based pharmacokinetic
PCA	Principal component analysis
PCs	Pericytes

PCL	Poly(ε-caprolactone)
PDMS	Poly(dimethylsiloxane)
PE	Primitive endoderm
PEG	Poly(ethylene glycol)
PEGDA	Polyethylene-glycol diacrylate
PEGDMA	Polyethylene-glycol dimethacrylate
PEG-(PTMC-A)$_2$	Poly(trimethylene carbonate)-b-poly(ethylene glycol)-b-poly(trimethylene carbonate)-acrylate
PEO	Poly(ethylene oxide)
PGA	Poly(glycolic acid)
PHEMA	Poly(hydroxyethyl methacrylate)
PIPPAAm	Poly-N(isopropylacrylamide)
POC	Point-of-care
POEGMA	Poly(oligo(ethylene glycol)methacrylate)
PLA	Poly(lactic acid)
PLGA	Poly(lactic-co-glycolic acid)
PLLA	Poly(L-lactic acid)
PMDA	Japan's Pharmaceuticals and Medical Devices Agency
PMMA	Poly(methacrylic acid)
PNIPAAm	Poly(N-isopropylacrylamide)
PVA	Poly(vinyl alcohol)
RBCs	Red blood cells
RGD	Tripeptide consisting of arginine, glycine, and aspartate
Red-Ox	Reduction–oxidation
RM	Regenerative medicine
RNA	Ribonucleic acid
RNase	Ribonuclease
ROCK	Rho-associated, coiled-coil containing protein kinase
ROS	Reactive oxygen species
RPTEC/TERT1	hTERT immortalized human renal proximal tubular epithelial
SDM	Synthetic differentiation medium
SDS	Sodium dodecyl sulfate
SDS-PAGE	Sodium dodecyl sulfate-polyacrylamide gel electrophoresis
SEM	Scanning electron microscopy
S-L PS	Solid–liquid phase separation
(s)SMCs	(Synthetic) smooth muscle cells
SP5D	Sphericalplate 5D
STL	Standard triangulation/tesselation language
STR	Stirred tank reactor
TE	Tissue engineering
TEBV	Tissue-engineered blood vessel

TEMED	Tetramethylethylenediamine
TERT	Telomerase reverse transcriptase
THF	Tetrahydrofuran
TIPS	Thermally induced phase separation
TGF	Transforming growth factor
TJ	Tight junctions
ULMP	Unfrozen liquid microphase
UV	Ultraviolet
VEGF	Vascular endothelial growth factor
Wt	Weight
2D	Two-dimensional
3D	Three-dimensional

Contents

Introduction to 3D Cell Culture

1

Dominik Egger and Sabrina Nebel

Contents

D. Egger (✉)
Institute of Cell and Tissue Culture Technologies, University of Natural Resources and Life Sciences, Vienna, Austria
e-mail: Dominik.egger@boku.ac.at

S. Nebel
University of Natural Resources and Life Sciences, Vienna, Austria
e-mail: sabrina.nebel@boku.ac.at

© Springer Nature Switzerland AG 2021
C. Kasper et al. (eds.), *Basic Concepts on 3D Cell Culture*, Learning Materials in Biosciences,
https://doi.org/10.1007/978-3-030-66749-8_1

What You Will Learn in This Chapter
This chapter explains the importance of 3D cell culture and highlights its potential and benefits in comparison to traditional 2D cultivation. Since 3D cell culture is supposed to mimic the in vivo situation, we will also have a close look on the complex composition of the in vivo microenvironment and why 2D cultivation does not resemble the in vivo situation. Furthermore, we will cover the most important approaches for 3D cell culture which are matrix-free and matrix-based cultivation as well as bioprinting. Also, the main applications of 3D cell culture, in vitro tissue or disease models and tissue engineered constructs for tissue repair or regeneration, will be introduced. Finally, you will learn that 3D cell culture has its limitations and challenges when it comes to the analysis and monitoring of 3D cell culture processes.

1.1   3D Cell Culture: The Bridge Between Bench and Bedside

The beginnings of Regenerative Medicine (RM) can be pinpointed back to the late 1970s where first discoveries on artificial skin paved the way to far more advanced research and commercialization of RM products. Today the field of RM comprises cell-based therapies, gene therapies, biologics and small molecules, tissue engineering approaches and utilization of stem cells for drug discovery, toxicity testing and disease modelling with the aim of replacing damaged tissue, activating the body's own healing response or delivering molecular and gene therapies to targets. Cell and tissue banks as well as service companies are making more than half of the involved enterprises. Regarding therapeutic related companies the largest group provides cell-based therapies followed by tissue engineering products. The paradigm "from bench to bedside" describes the principle that findings from basic research are translated into therapeutic clinical applications. However, there is an obvious and serious gap between bench and bedside in all fields of RM. As a result, Fernandez-Moure proposed that the current paradigm "from bench to bedside" needs to shift to "from the bedside to bench and back" [1], meaning that not only findings from basic research should be the basis for the development of new therapies but also the clinical needs drive the focus of basic research. So, how can this be realized?

Five drivers for research in RM have been identified: (1) manufacturing, (2) reimbursement, (3) multicompetence collaborations, (4) regulatory compliance and (5) clinical trials [1]. Although manufacturing and reimbursement issues are critical, they are more relevant in later stages of product development and not of primary concern at the level of basic or applied research. Multicompetence collaborations are valuable but need to be elaborated and strengthened before research is actually conducted. However, regulatory compliance and clinical trials are issues that researchers should consider from the very beginning. After the generation of basic or pre-clinical data, many ideas or approaches for novel products or therapies in RM cannot be pursued further since they are conceptually incompatible with the approval for clinical trials. National regulations and clinical trials are very specific

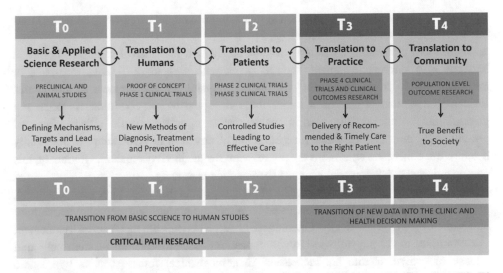

Fig. 1.1 Translational research comprises five different phases depicted as T0–T4 where T0–T2 comprises the translation from basic research to clinical trials and T3/T4 comprises follow-up studies. Each phase is defined by distinct challenges and moreover need to communicate with the prior and advancing phase; modified from [3]

challenges of RM. To avoid those so-called valleys of death where the development of novel therapies comes to halt, prior considerations regarding the national regulations at the basic and applied research level are crucial. Researchers and investors need to clarify in advance whether there is a true clinical need to conduct the following steps in research and translation and early classification of the product will help to make considerations on the requirements of the later approval and manufacturing process. Therefore, considerations on the basic research level may improve the later translation process.

The translation of results from basic research to clinical application is a great challenge. In general, about 85% of therapies fail in phase I or II clinical trials and only half of products that enter phase III trials are then approved [2]. Besides the fact that some products are just not effective in a statistically significant number of human beings, possible sources of failure during clinical trial but also basic research exist. Translational research comprises at least five phases with four "translations" in between and all of them need to be successful for the successful development of a novel therapy (Fig. 1.1). Here, researchers, clinicians, health care institutions and industry must work hand in hand which requires that all involved parties "speak the same language". Especially in the area of advanced therapy medicinal products (ATMPs), several uncertainties exist. Therapeutically positive effects observed in the lab are often not present in first human trials and thus projects come to a halt at early stages. To increase the predictability of pre-clinical studies or refine potency testing of ATMPs and biologicals, proper models for in vitro testing are of tremendous importance. These models must be relevant from a regulatory perspective and robust with regard to safety, efficacy, purity and dose response. Owing to advances in 3D cell cultivation and dynamic cultivation systems, in vitro models are getting better and thus more predictable for clinical outcome [4, 5].

Traditional cultivation conditions like the cultivation of adherent cells on 2D plastic surfaces in a static environment, such as given with a standard T-flask, well-plate or Petridish, are far from representing the natural, physiologic environment of these cells. Probably, data that is generated under these non-physiologic conditions is often not physiologically relevant. In fact, a major reason why therapeutic effects observed pre-clinically under 2D conditions are often not observed in in vivo studies is potentially because the cultivation conditions are not relevant for the in vivo situation. Obviously, this compromises the development of novel therapeutics massively.

Research from the last decade proved without any doubt that when biological, chemical, physical and mechanical cues are adjusted to mimic the physiologic environment, cellular behaviour changes dramatically. Often, these observed effects are therapeutically relevant. For example, under physiologic oxygen concentrations, proliferation of stem cells and release of growth factors are increased. Also, mechanical forces such as hydrostatic pressure or shear stress in a physiologic range induce the differentiation into specific lineages.

The biological context comprises cell–cell contacts, cell–extracellular matrix (ECM) contacts as well as the ECM itself. Chemical cues embrace growth factors, cytokines, nutrients, salts and toxic compounds. Physical cues are temperature, partial gas pressures or viscosity and mechanical cues describe the environment generated by physical forces, such as shear, pressure or tension. To mimic the physiologic environment myriads of 3D matrices from different materials and of different shape and geometry have been developed together with various bioreactor systems for the application of mechanical forces. The implementation of physiologic conditions is expected to increase the predictability of in vitro testing for in vivo trials.

Expanding the cellular in vitro environment by a third dimension adds immensely to the generation of a physiologic environment. To extend cellular growth to the third dimension, supportive structures, called matrices or scaffolds, have been engineered from numerous materials (see Chaps. 2, 4, 5). However, the vast amount consist of ceramics (like tri-calcium phosphate or hydroxyapatite), synthetic polymers (like polystyrene, poly-L-lactic acid or polyglycolic acid) or natural polymers (such as collagen, alginate or silk), each having different physicochemical properties, architecture and biodegradability [6]. The material properties affect the cellular behaviour in many ways. Obviously, every material has its own advantages and disadvantages and must be, therefore, chosen to fit the respective biological requirements.

In conclusion, the main reasons for 3D cell culture is the generation of a physiologic environment to generate physiologically relevant data to foster the development of better in vitro models for, i.e., drug testing and tissue engineered constructs to increase the outcome of translational research.

Questions
1. Why is 3D cell culture important in translational research?
2. What is the most important advantage of 3D cell culture?
3. Why does the development of many products and therapies stop after the first clinical trial?

1.2   The Physiologic Microenvironment of Cells

1.2.1   Cells Shape Their Environment and the Environment Shapes the Cells

To understand the importance of switching from 2D to 3D culture systems we need to comprehend the natural environment of cells within a living organism. When we are first introduced to cell biology, we encounter a very generic image of a cell, its organelles, their function and interactions within the boundaries of the outer membrane. Further on, different cell types and their specialized functions are studied. It is now very important to remember that in vivo tissue consists of a highly specialized arrangement of different cells and that this architecture is a crucial point for tissue function. The skin serves as an excellent example, the epidermis as outermost layer consisting of very tightly bound keratinocytes to create a barrier to the outside. Underneath we find the basement membrane for stabilization and the dermis, harbouring blood vessels, hair follicles and sensory functions. Furthermore, on the example of structural tissues like tendon and bone we can see that a lot of non-cellular material not only is present but actually contribute to tissue functionality.

A fascinating point to consider when talking about the cellular microenvironment is that not only function follows form but also form follows function, meaning that within the complex system of a living organism there is a constant exchange of information from the cellular level to the three-dimensional structure of tissue. The different cell types create the tissue architecture by assembling themselves in distinct layers or zones, constantly remodel their surrounding space by depositing or degrading extracellular matrix molecules. On the other hand, the surrounding ECM and the mechanical, physical and biological cues within the 3D structure influence cell behaviour and differentiation [7].

1.2.2   Components of the Cellular Environment

How exactly does this environment look like? The space outside a cell can be inhabited by other cells, ECM, hormones and other solutes, all in an aqueous environment. Cell-dense tissues are, for example, found in the brain and the spinal cord, but in a lot of regions of the body ECM occupies most of the extracellular space, rather than the cells. This ECM is made of an intricate network of proteins and polysaccharides secreted mostly by a special

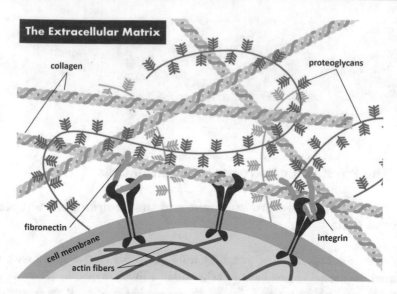

Fig. 1.2 Schematic illustration of the ECM; the main components of the ECM include fibrous proteins (collagens, fibronectin) and large polysaccharide chains linked to core proteins (proteoglycans), the cells cytoskeleton (actin fibres) is connected to these ECM macromolecules via transmembrane proteins (integrins), especially fibronectin plays an important role in cell-to-ECM and ECM-to-ECM binding

family of cells, the fibroblasts. Two large classes of macromolecules can be distinguished as building blocks of the ECM (see Fig. 1.2). Large polysaccharides, the so-called glycosaminoglycans (GAGs) that are usually covalently bound to proteins. These proteoglycans (combination of the terms protein and glycan) form highly hydrated gels, in which the second class of macromolecules, fibrous proteins, can be embedded in. Plenty of carboxyl and sulphate groups running along the polysaccharide chains cause a negative charge that attracts cations (especially Na+). The resulting osmotic pressure causes an immense amount of water to be drawn into the matrix, ensuring a high resistance against compressive forces. The fibrous proteins give further strength to the hydrogels by providing robustness under tension and are necessary for cell anchorage within the ECM. With differing proportions of matrix macromolecules and ways in which they are organized, a great variety of structures can be built, ranging from rigid, hard ECM in bone to the extreme elasticity of blood vessels or the rope-like, tension-resistant structure of tendons [8].

GAGs are unbranched and composed of repeating disaccharide units with one unit always being an amino sugar. Hyaluronan, chondroitin sulphate, heparan sulphate and keratan are the four main groups. As already mentioned, they are usually bound to a core protein (with hyaluronan being the exception) forming proteoglycans. These macromolecules are huge compared to other glycoproteins and can even further assemble to highly organized aggregates.

From the fibrous proteins, collagen is by far the most abundant one comprising 25% of total protein in mammals. Collagen has a hierarchical structure starting with three

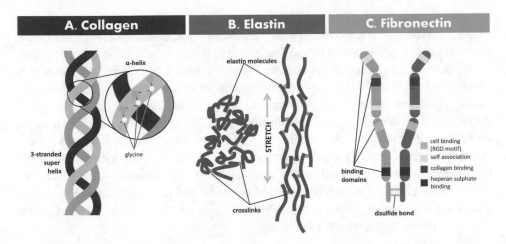

Fig. 1.3 Schematic representation of the ECM proteins collagen, elastin and fibronectin. (**a**) Collagen is composed of three alpha helices that form a triple stranded super helix. This is made possible by the repetitive sequence with every third amino acid being a glycine, which has only a hydrogen atom as a side chain. (**b**) Elastin fibres are formed by covalent crosslinking of single elastin molecules that curl together in the relaxed state and straighten out once stretched. This allows for high elasticity within the tissue. (**c**) Fibronectin is a dimeric protein with specific domains for self-association, binding of other matrix molecules and especially cell attachment. The RGD motif in those domains can be bound by integrins

polypeptide chains in an alpha-helix that are wound around each other in a helix again, a so-called triple superhelix. This is possible as the peptide chain is rich in proline and glycine. Every third amino acid being glycine, with only a hydrogen atom as side chain, thus the three strands can form this tight contact (see Fig. 1.3a). With 25 different alpha-chains encoded in the genome, there are a lot of possibilities, yet only around 20 have been identified so far. The most found ones are type I, II, III, V and XI. Multiple of these collagen superhelices together can form a collagen fibril by regular stacking of molecules on top of each other once secreted and strengthened by covalent crosslinking. The regularity of the stacking can even be seen in the electron microscope as cross-striations of a fibril every 67 nm. Multiple of such fibrils often form even thicker and rope-like bundles: collagen fibres. Alignment and crosslinking of these ECM macromolecules is especially important in load-bearing tissues where high tensile stress is present. Other types do not assemble into fibres; however, some are thought to link the fibrils together (IX, XII). Others can form networks (IV, VII) and are essential for the formation of the basal laminae [9].

Next to resisting strong tensile or compressive forces a crucial parameter of the ECM in many tissues is also flexibility, with the most prominent examples being blood vessels, lungs and skin. This is achieved by an extensive network of elastic fibres. Interwoven with varying degrees of sturdy collagen fibres they control the elasticity and resilience of the tissue. Elastin, as the name implies, is one of the main components of elastic fibres. It is similar to collagen regarding the amounts of proline and glycine; however, the molecules do not assemble into helices and fibrils but take on a random coil conformation (see Fig. 1.3b). These coils are covalently bound to each other, thus allowing the network to be stretched and recoil back to the original shape [10].

While collagen and elastin are essential for giving tissues their mechanical properties, fibronectin is of special importance for cell attachment to the matrix. Studies have shown that blocking fibronectin during embryonic development inhibits gastrulation [11]. It is a dimeric glycoprotein that has distinct domains for both ECM molecules and cells to attach to. ECM-binding domains allow binding to other fibronectins, collagen, fibrin and heparan sulphate, whereas the RGD sequence (Arg-Gly-Asp) in the cell binding domain is what is recognized and bound to by cell adhesion molecules (CAMs). Similarly, the RGD motif can be found also in laminin, a heterotrimeric cross-shaped protein. It is the main component of the basal lamina, because of its ability to form sheets. This is possible due to 3 of the 4 arms are prone to bind to other laminins [12].

1.2.3   Stimuli Within the Cellular Environment

A cell can receive environmental cues from its surroundings are based on chemical and physical phenomena, highly intertwined with change of one parameter influencing dozen others. Still we want to try to divide stimuli acting on the cell into four groups, which will help to better understand the cellular microenvironment: biological, chemical, physical and mechanical cues (Fig. 1.4).

To begin with we will take a short look on what we classified as physical stimuli. One very important factor regarding this is temperature, an atmosphere containing oxygen and the presence of water. Within an organism temperature is tightly controlled within a rather small range. In mammals the optimal internal temperature is at 37 °C, decreasing temperature is counteracted with, for example, muscle contractions simply known as shivering. In the case of an infection the body can also harness the higher sensibility to heat of the invading pathogens by regulating the temperature up resulting in a fever. On an intracellular scale we must take even more into account that temperature is in fact particle movement. With increasing temperature particle speed increases, which in this environment has influence on the speed and efficiency of enzymatic and chemical reactions. Next to temperature, physiological atmosphere has to be present, in the context of cell culture this is especially the partial gas pressure of oxygen and carbon dioxide. The concentration of the dissolved gases in the cellular environment is crucial for cell survival. Oxygen is needed for cellular respiration, the chemical reaction in which simplified glucose and O_2 is converted into water, CO_2 and energy, which is partly stored as ATP, partly lost as heat. Important to consider is that in different parts of an organism different oxygen concentrations are present, leading to a distinct demand of different cell types [13]. The amount of oxygen present can even be a cue for stem cells during development to push them into a specific differentiation lineage. Lastly, as the cellular environment is aqueous but houses myriads of macromolecules, there are deviations from the fluid dynamics of pure water. Viscosity can influence the diffusion rates into the gaps between substrate and

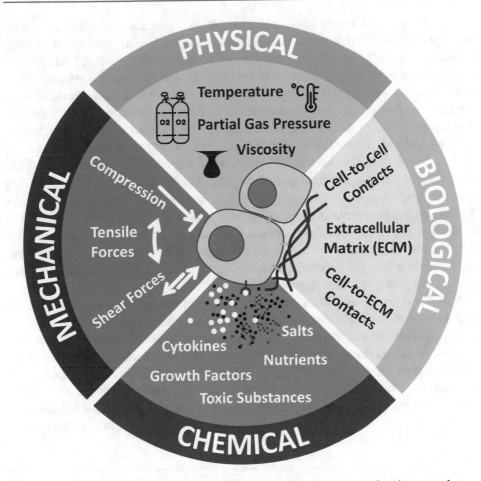

Fig. 1.4 Overview of the different stimuli acting on a cell. The major types of environmental cues a cell receives can be divided into physical, chemical, mechanical and biological stimuli

cell membrane or in extreme cases even restrict cell movement [14]. These three factors are mostly regulated systemically to allow for ambient conditions in the cell microenvironment rather than be established directly within it. This, as we will discuss later in Sect. 3.1, makes external control of these factors' prerequisites for ex vivo maintenance of cells.

The second kind of cues the cell receives is of enormous complexity and variety. Chemical stimuli are extremely diverse and range from ions/salts to nutrients, cytokines and growth factors and are found in high abundance in the extracellular space. However, concentrations inside and outside of the cell must be balanced. Due to osmotic pressure, cell swelling or shrinking may cause fatal damage. Although this balance is persistent, concentration gradients of one kind of solute are necessary for a lot of cellular mechanisms. Whether by passive transport down the concentration gradient or active transport against the gradient, especially ions drive important mechanisms, such as muscle contraction, nerve conduction, hormone secretion and sensory processes. Next to ions, cells need nutrients in the form of amino acids, as building blocks of proteins. Especially important is the supply with amino acids that cannot be produced by the organism itself, hence the name essential amino acids. Furthermore, carbohydrate, the main source of energy being glucose, fatty acids and lipids for the formation of lipid bilayers, and vitamins and essential amino acids which cannot be synthesized by the cells but are essential for growth, are needed. For some enzyme functionality trace elements like copper zinc and selenium are also necessary. Lastly, peptides and proteins that again are a rather heterogeneous group. Of great interest are cytokines and growth factors, peptides and proteins that most notably initiate and regulate the cell cycle, but also are mediators of immune reactions, chemotaxis, growth and differentiation. Some examples are interferons (IFs), interleukins (ILs), colony stimulating factors (CSFs), chemokines, fibroblast growth factors (FGFs), transforming growth factors (TGFs) [15]. All these proteins are signalling molecules that are either secreted by cells into the environment or are within the membrane. Cells with a matching receptor can respond to these signals. Thus, short-range and long-range signalling between cells is made possible. The supply and foremost the composition of all the compounds in the cellular microenvironment can greatly influence the behaviour of a cell.

A different source of cellular stress can arise from mechanical forces on the cells. Excessive forces can cause material failure in a cellular context; this would mean rupture of the cell membrane and ultimately cell death. However, below this upper limit, mechanical stress is an important environmental stimulus. When we are talking about mechanical forces, what we mean are physical forces that have a magnitude and most importantly a direction, so called vectors. If we break them down to the most basic forms, we can have compression, tension, bending and torsion [16]. More complex forms include hydrostatic pressure and shear stress, as depicted in Fig. 1.5. Especially tissues of the skeletal system

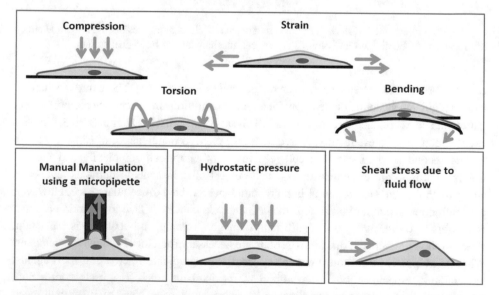

Fig. 1.5 Types of mechanical forces acting on a cell; blue arrows indicate the direction of the force

have to withstand and are moulded by mechanical forces making them an essential component of the cellular environment.

As mentioned above, the three-dimensional architecture, organization and composition of tissues is what makes them functional. Mechanical stress can guide the cells during differentiation or proliferation and trigger secretion of different ECM molecules. They also help cells to organize spatially, to align like we see in muscle and tendon or build layered structures for, e.g., skin. If we think about this, the question arises how exactly do the cells "feel" mechanical stress put on them directly or on the ECM around them? This is where mechanotransduction comes into play. Mechanotransduction refers to the translation of an external physical force to a biological intracellular signal that can be processed by the cell. The mechanisms used for this is discussed in the last group of outside cues—biological stimuli [8].

Although biological stimuli can be broken down to chemical, mechanical and physical phenomena, their specificity and importance in the cell microenvironment is substantial enough to highlight them as a separate group. Mechanical stimulations are converted to biological cues by specified structural components that we will discuss in detail.

There are several different cell-to-cell contacts that each is designed for specific purposes. For example, as mentioned above, cells frequently must create a barrier between two regions in the organism or to the exterior. To achieve such a sealing function, multiprotein complex the so-called tight junctions (TJ) are needed. They consist of strands of occludin and claudin, which bind to each other outside of the cell and anchor to intracellular actin. The composition of TJs regulates what solutes can pass through and which ones are excluded. In addition to providing mechanical stability and a barrier, they restrict movement of membrane components, thus are vital for preservation of cell polarity.

Adherens junctions (AJ) but also desmosomes serve as mechanical linkage between cells. Adherens junctions are composed of dimeric transmembrane proteins called cadherins that act as a link between the actin cytoskeleton between two cells. The extracellular domains bind to those of another cell, whereas intracellular domains interact with catenins that than bind to the actin fibres. Cadherins comprise also the extracellular binding domains of the second type of adhering junctions the desmosomes. They can be compared to a rivet, with dense disc-shaped desmoplakin plates as linkers between the intracellular domains of the cadherins and keratin or desmin filaments. They are found in tissue types that must withstand intense mechanical forces as they are one of the stronger cell adhesion types [17].

The last type of cell-to-cell contacts is less important for mechanical stability but serves in cell communication. Gap junctions are clusters of cell-to-cell channels that directly connect the cytoplasm of both cells. Each channel is made from two hemichannels called connexons, one per cell. This enables direct communication of the cells by diffusion of ions, second messengers or small metabolites. This is used in low vascularized tissue to improve nutrient distribution. They furthermore act as electrical synapses in neurons and in the heart [18].

The environment of a cell is seldomly composed of cells entirely, as we learned earlier. Next to cell connections they also need to be anchored to the surrounding ECM.

Specialized for the connection of epithelial cells to the basal lamina are hemidesmosomes. They are closely related to desmosomes. Intracellularly they look the same as desmosomes, connecting to the intermediate filaments of the cytoskeleton. For extracellular adhesion, integrins are utilized that bind to the laminin proteins of the basal lamina. Integrins are heterodimeric transmembrane glycoproteins, their extracellular domains have binding sites for an RGD sequence which we learnt is present, for example, in fibronectin, vitronectin and laminin.

Integrins are also the main player in another form of cell-to-matrix adhesion, the so-called focal adhesions. Those are large macromolecular complexes that connect the actin filaments of the cytoskeleton to ECM molecules. On the outside they link to ECM molecules, whereas on their intracellular domains they are connected to the actin filaments by adaptor proteins. Next to anchoring, presence of signalling proteins like tyrosine kinases and cSrc (cellular Src; pronounced "sarc", as it is short for sarcoma) and focal adhesion kinase (FAK) indicates the secondary role of focal adhesions—signal transduction. Constant association and dissociation of proteins represent fundamental mechanisms for the transmission of signals from and to other parts of the cell. Focal adhesions can persist once they are formed for stationary cells. For cell movement, however, new ones have to develop, and old ones have to be disassembled for the cell to "crawl" forward [8].

Questions
4. What types of fibrous proteins can be found in the extracellular matrix and what is their main function?
5. What main types of cues does a cell receive in its microenvironment?
6. Which types of cell-to-cell contacts are there and what is their function?

1.3   Standard Cell Culture Vs. 3D Cell Culture

1.3.1   Cell Culture Basics

First maintenance of cells outside of living organisms was accomplished already in the early twentieth century [19]. From the 1950s on, the technological development made in vitro cell culture easy and reproducible, and the fundamental procedures from then did not change much over time. Cells are taken from their native tissue and placed into a foreign environment that tries to emulate the physiological conditions, ranging from minimum requirements to more elaborate models. Cultures are kept in incubators to control temperature, which is set to a constant 37 °C for most cultures, the humidity to minimize evaporation, and the carbon dioxide content of the gas phase (5%). This is important to stabilize the pH of the cultures, by interaction of the CO_2 with the mostly carbonate-based buffering systems within cell culture medium. Next to its function to keep the optimal pH of 7.4, cell culture medium provides the cells with basic components like glucose, inorganic salts, amino acids and vitamins. There are different formulations for different cell types of this basal medium which is often supplemented with serum from bovine origin. This adds a rich but non-defined mixture of hormones growth factors and attachment factors. For reduction or replacement of bovine calf serum use due to ethical reasons, researchers are switching to alternatives like human platelet lysate (waste product made from expired platelet concentrate) or chemically defined medium. Many culture media also contain phenol red to indicate the pH or antibiotics and antimycotics to reduce the risk of microbial contamination. The cells are maintained within this liquid either in suspension or for the majority grown on a solid substrate as the majority of cell types are adherent. Typical types of tissue culture vessels are Petri dishes, flasks and well plates where the cells are attached as a monolayer on the bottom and covered with medium. Upon complete coverage of the culture area, cells are detached using enzymes (e.g. trypsin) and reseeded into multiple fresh dishes for further cultivation [19, 20].

1.3.2 Cells on a Plane Surface Versus Cells in (3D) Space

As stated earlier, flasks and dishes are the standard for adherent cells for the last decades. Initially vessels were made of glass, but by the 1960s plastic culture vessels became available [15, 19]. Polystyrene, a long carbon polymer with benzene rings attached to every other carbon, seems to be a logical choice because of its optical clarity, easy fabrication and can be sterilized by irradiation. Unfortunately, there is one major obstacle: the surface of polystyrene is highly hydrophobic; thus, cells can hardly adhere to it. Therefore, surface treatments are necessary to create a more hydrophilic environments where fibronectin and vitronectin contained in the medium can adhere to and ultimately the cells. High energy oxygen ions are generated during the treatment which oxidize the polystyrene surface. Once in contact with medium the surface becomes negatively charged. The oxygen ions can either be generated under atmospheric conditions by corona discharge or under vacuum by gas-plasma [21]. Rapidly, these now tissue-culture compatible, single-use plastic culture vessels became standard in laboratories, replacing glassware which needed extensive cleaning procedures.

Despite all the advantages, the mechanical properties of polystyrene and glass differ largely from what cells in vivo are used to. Polystyrene, for example, has a stiffness of 1–3 GPa, which is about six orders of magnitude higher than the majority of tissues has [22]. In many studies the effect of substrate stiffness on different cell types and mechanism has already been proven [23–27], including drug sensitivity, proliferation, gene expression and differentiation.

Now, the surface rigidity and chemistry are not the only problem to be tackled in traditional cell culture. Substrate surface coatings with ECM molecules or artificial materials to resemble stiffness more closely are an improvement, but the cells still experience a non-physiologic spatial environment. The shape of a cell is tied to whether it can form adhesion sites all over its surface or only on one side. As cells are grown on a planar substrate in traditional culture, they can only adhere on their dorsal side and spread almost unhindered in the x-y-plane, resulting in a very flat cellular body and a forced polarity. Cell polarity in general stands for the asymmetric distribution of cell membrane and other organelles within a cell and is important during cell movement and in specialized cell types (e.g. barrier function of epithelial cells). However, the constant forced polarity in usually non-polar cells can lead to changes in cell function. For example, a different surface-to-volume ratio of a cell can influence levels of signalling initiated by cell surface receptors [28]. For example, FAK signalling, or more specific FAK autophosphorylation was down-regulated in 3D-cultures compared to two-dimensional controls [29]. Differentiation, proliferation and apoptosis could all be shown to be affected by the geometry of the cell shape [30].

Not only lack of mechanical cues and abnormal cell-to-cell interactions are problems in classical culture, also chemical and physical parameters are different from the in vivo situation [31].

As mentioned before, inside tissues there are gradients of oxygen and nutrients, concentration depending on the distance to blood vessels and permeability of the surrounding ECM. In the 2D set-up however, cells have almost unlimited access to nutrients, signal molecules and metabolites inside the culture medium. Atmospheric oxygen levels in incubators are even higher than anywhere within the body. This overexposure can result in formation of reactive oxygen species (ROS) and in the end in cell damage. Even if there is no damage, this unnatural surplus of nutrients and O_2 changes the cellular behaviour.

Transitioning from the 2D flat surfaces and expanding the culture with a z-axis can greatly improve native-like cell behaviour (Fig. 1.6). Generally, there are two possibilities to achieve this and both take advantage of the fact that adherent cells can form connections to other cells, to ECM molecules and can produce their own ECM molecules. Either cells only are aggregated or seeded into an artificial ECM in form of sponges or hydrogels. These substrates can be made from natural ECM components, various biomaterials, ceramics or synthetic polymers, we will go into detail later. Now first of all, the main differences in three-dimensional cell culture compared to traditional culture shall be discussed. The most important factor is that cells have the ability to form adhesion sites all over their surface, whether to other cells or the biomaterial. This eradicates the artificial apical polarity and results in a non-polarized morphology as for most cell types seen in vivo.

Moreover, cells are more restricted in 3D than on normal culture dishes. This might sound like a drawback initially but considering the natural microenvironment cells reside in, for them to migrate they have to interact with the ECM by degrading adhering and reassembling it. This is also true for cells in in vitro 3D cultures. Also, the properties of the used material can be used to direct cell movement. Small pore sizes within a sponge can restrict cell movement but still allow for nutrient and waste diffusion while hydrogels made

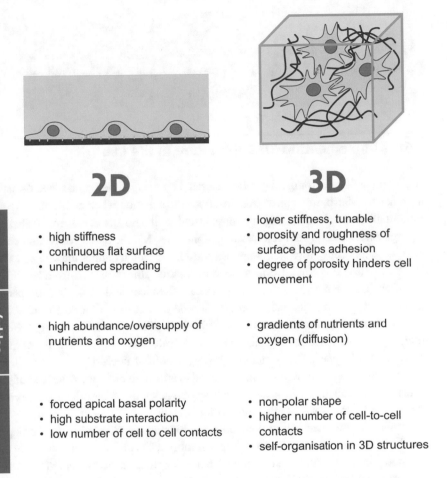

Fig. 1.6 Comparison between standard 2D cell culture and 3D cell culture. Schematic depictions of the cell shape (grey), the substrate (dark blue) and cell adhesion points (yellow) show drastic differences between standard 2D culture and three-dimensional cell culture

of non-degradable biomaterials can entrap cells for safe and continuous delivery of secretory factors [32].

Another important difference is that oxygen, nutrients, signalling molecules and waste products are now restricted by diffusion rates. The resulting gradients from graft surface to its centre play an important role in physiological and pathological processes. Especially in cancer research, 3D models have proven to be more predicting of treatment outcome than the previous used 2D cultures. This field is a pioneer of 3D cell culture, however continuous advances in manufacturing technologies open new possibilities for the remaining scientific community.

Questions
7. What are the basic requirements for cultivating adherent cells in vitro?
8. What are the main differences between 2D and 3D cell culture in respect to the shape and geometry of the cells?

1.4   Implementing 3D Cell Culture in the Lab

In general, a 3D environment can be generated by using either matrix-free or matrix-based approaches. Matrix-free cultivation describes the situation when cells adhere to each other without an external supporting structure. Usually, this results in spheroidal shaped cellular aggregates, also frequently referred to as spheroids. Matrix-based cultivation describes the situation when cells grow on a supporting structure, which defines the size and shape of the final cell-matrix construct. Besides the term "matrix", the terms "scaffold" or "biomaterial" are often used synonymously. The physical, chemical and biological properties of the matrix affect the cellular behaviour. Thus, the matrix can be used to drive the cellular behaviour into a specific direction (i.e. differentiation into a specific lineage). However, the interplay between matrix and cells needs to be evaluated for every specific combination of matrix and cell type in the context of the application of interest.

How can adherent growing cells be forced to adhere to each other, instead of growing on a surface? The methods to generate cell aggregates will be described only briefly in this chapter, as Chap. 7 covers this field in more detail.

In general, the methods to generate spheroids are subdivided in two groups: cluster-based self-assembly and collision-based assembly. Cluster-based self-assembly takes place in a static environment, i.e. a cell culture plate. Single cells are prevented from attaching to a surface and thus come in contact to form aggregates. Collision-bases assembly needs a dynamic environment where cells collide by mixing, centrifugation of the cell suspension. The most widely used approach to generate spheroids is the hanging drop culture: a drop of cell suspension is pipetted on a surface (i.e. a Petri dish) and then the dish is turned upside-down. This results in a hanging drop in which the cells sink down by gravitational force and form a spheroid because they cannot grow on the inner surface of the liquid drop.

The next widely used approach is the use of cell-repellent surfaces, also termed as ultra-low attachment, no attachment or ultra-low adhesive surfaces. These surfaces are made from hydrophobic materials or are nanostructured in a way that cells cannot attach to it. Both forces the cells to adhere to each other. Many cultivation vessels are now available with cell-repellent surfaces.

Further, technological advancements made it possible to manufacture cavities in the range of only several micrometres. Now, vessels with hundreds or thousands of microcavities on the surface are available. The cavities are manufactured in a way that, theoretically all cells sink down in of these cavities after seeding. This allows for an almost

100% incorporation of cells and homogeneously shaped and sized spheroids. A more detailed description of this method and commercially available equipment can be found in Chap. 7.

Other methods, such as microfluidics, thermal responsive surfaces or magnetic force have also been demonstrated to generate cell aggregates, but most of them afford either specialized equipment or are not commercially available.

Overall, the generation of spheroids has become very easy and a lot of commercially available labware is available. Thousands of homogeneous spheroids can be generated in little time by using automated pipetting robots. Thus, spheroid cultivation is used for high-throughput applications, i.e. in drug testing. However, some drawbacks exist. Spheroids are limited in size due to diffusion limitation (which will be discussed later in this chapter). Depending on the system that is used, spheroids are often inhomogeneous in size and shape. For example, cell-repellent 96-well round bottom well plates are widely used to generate aggregates. However, after seeding, the cells often several smaller spheroids instead of one aggregate that comprises all cells. This again can lead to false results quickly. Furthermore, medium change can easily result in losing the spheroid. Thus, long-term cultivation is risky.

Besides matrix-free, also matrix-based methods can be used to generate 3D cultures. In the following, we will give an overview on the most important classes of scaffolds which are described in detail in Chaps. 4 and 5. Synthetic scaffolds are subdivided into polymers, ceramics, metals and composites of these. The manufacturing process is referred to as "scaffolding" and comprises countless procedures and methods. Biological or natural scaffolds are derived from living organisms and then further processed. In most cases these are decellularized and dehydrated tissues or preparations from these, i.e. polymers such as collagen or fibrin. There is no such thing as the "perfect scaffold" that is suitable for every kind of application. Instead, the scaffold needs to be chosen according to the application. In general, the material should be chosen whose properties resemble the in vivo situation the best. Inherent material characteristics such as porosity, pore size and distribution, surface to-volume ratio, mechanical characteristics and surface chemistry have an influence on cellular behaviour. Cell attachment, migration, proliferation and differentiation were shown to be impacted by these material characteristics. In return, 3D cultivation has a remarkable impact on the cell shape, cell–cell and cell–ECM interactions and thus heavily affects the outcome of drug testing, for example [26, 33].

A third possibility to generate 3D cell cultures should be mentioned here separately, since it cannot be classified into matrix-free or matrix-based cultivation. The rapid development in the field of additive manufacturing, also referred to as 3D printing, made it possible to arrange cells in a three-dimensional manner. This can be achieved by inkjet, micro extrusion or laser-assisted bioprinting. All these methods will be discussed in Chap. 8 in detail. In general, a cell suspension is prepared in a printable carrier material, a natural or synthetic polymer, and printed in a defined manner in x-, y- and z-dimension, building up a construct with cells and the carrier material. If the architecture, necessary cell types and ECM of the target tissue are known, it is theoretically possible to build up this tissue from the scratch, including even the vasculature. With this technology researchers hope to be able to print not only small tissues but entire organs, in the future. As appealing as this approach sounds, the reality is much more complex and challenging.

Questions
9. What are the three main approaches to generate 3D cultures?
10. What are the most common techniques to generate spheroids?
11. What are the advantages of matrix-free cultivation?
12. What are the advantages of matrix-based cultivation?

1.5   Applications

As mentioned in the introduction, 3D cell culture is mainly supposed to bridge the gap between "bench to bedside" in translational research. Thus, it can foster the development of new therapies. The main application of 3D cell culture is the generation of more relevant in vitro models. This includes models of specific healthy but also diseased tissues. Second, in vitro engineered tissues are also used for tissue repair or regeneration in regenerative medicine, meaning the matrix-free aggregates or matrix-cell constructs are supposed to be transplanted to a patient. Both applications have in common that the in vitro generated tissue should be as close to the physiologic in vivo situation as possible. The use of 3D cell culture for in vitro models is not only interesting because it results in more relevant data but also reduces the need for either patient-derived samples for in vitro studies or in vivo animal testing.

In vitro test models are currently the most important application for 3D cell culture because in vitro models are used for drug development which has (currently) a much higher economic importance than tissue engineered products. Further, in vitro test models are used in research to study the development of tissues or diseases. Prominent examples for 3D in tissue models that are used to study drug metabolism and toxicity are healthy or diseased liver, intestine and brain models and a variety of tumour models [34]. In the field of drug development, animal testing is still required before entering the first-in-human trials. Often the therapeutically relevant effect of drugs is first observed in standard 2D cultures which are cheap but have a very low relevance. 2D culture often comprises only one or few cell types, thus systemic responses simply cannot be modelled. Thus, studies in animals have a higher relevance although it is clear that the translation of the results to the human organism is limited. When it comes to planning of animal test, researchers are required to justify the need for animal testing and explain why there are not alternatives. They must stick to the so-called 3 R rule: replace, reduce and refine. In the first place, they need to replace animal testing wherever possible. Second, they need to reduce the number of required animals to the necessary minimum while obtaining the same amount of information. And third, they need to refine the experimental procedures to keep animal suffering at a minimum. But even if the "Three R rule" is followed strictly, animal testing remains ethically questionable, costly and the translation of the results from animal testing to the human organism is still limited. Thus, 3D in vitro models are supposed to bridge the gap from preliminary 2D cell culture which are cheap but not relevant, to animal testing which is very costly but more relevant. 3D models have the potential to at least reduce the number of animal testing with the additional benefit of providing a more relevant "reaction" when using human cells.

Currently, the most widely used 3D models are matrix-free (spheroids) which are often referred to as "organoids", with only one or a few cell types [34]. This can be either cell lines, primary stem or tissue-specific cells or tissue-specific cells derived from embryonic stem cells (ESCs) or induced pluripotent stem cells (iPSCs). These organoids enable for high-throughput screening of drugs in automated robotic cell culture platforms. Further, they are often used in microfluidic "organ-on-a-chip" devices where the organoids are cultivated in a highly defined microfluidic system that mimics the organ environment. In these chips several organoids can be combined to mimic the interplay between several organs or tissues [35, 36]. More on this topic can be found in Chaps. 6, 7 and 8.

As mentioned before, 3D cell culture is also used to generate specific tissues for the replacement of damaged tissue in a patient. This traditionally called "Tissue Engineering". In general, a Tissue Engineering process starts with the isolation of primary tissue-specific (i.e. chondrocytes) or stem cells (i.e. mesenchymal stem cells) from the patient. The cells are then expanded in vitro to achieve the necessary cell number and the cells are seeded on a suitable matrix. The matrix-cell construct is then often cultivated in a bioreactor to foster cell proliferation on the matrix and support or induce differentiation into a specific lineage, if necessary. For this, it induces mechanical cues to mimic the physiologic environment and ensures constant nutrient supply and waste removal. After the matrix-cell construct has matured enough, it goes back to the patient to replace the injured tissue (Fig. 1.7).

Although, researches have tried to generate virtually every tissue in vitro by Tissue Engineering, only around 20 therapies are currently approved through the necessary scientifically-based, regulatory approval process, including in-human clinical trials overseen by internationally-recognized regulatory agencies, such as the U.S. Food and Drug Administration (FDA); the EU's European Medicines Agency (EMA); Japan's Pharmaceuticals and Medical Devices Agency (PMDA); South Korea's Ministry of Food and Drug Safety (MFDS), among others. Most of the products are approved for the treatment of damaged skin, cartilage, bone or cardiac tissue which is actually only a small number of target tissues.

In fact, the expectations that were raised by research in the field of tissue engineering were not fully met. The reason for that is that tissue engineering process comprises many steps that need to be perfectly orchestrated to ensure continuous safety and efficacy of the product. For now, variances in raw products, i.e. cells and biomaterials, are often too high and the process still contains too much manual handling to meet the high standards of

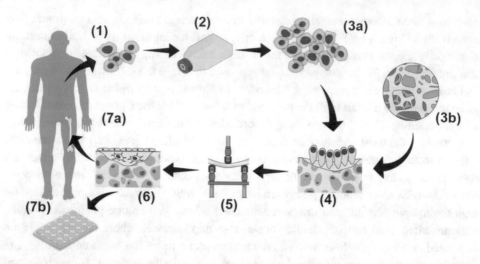

Fig. 1.7 Concept of the tissue engineering process: Cells are collected from a healthy donor site (1) and expanded in standard 2D culture (2). The propagated cells (3a) are combined with a biomaterial (3b) to create a 3D cell culture (4). Optional (mechanical) stimulation (5) can be applied to achieve tissue maturation. The final 3D cell culture construct (6) can then be used for clinical applications (7a) or as an in vitro test model (7b)

medicinal products required by the authorities such as the FDA or EMA. However, one prominent example of a tissue engineered product that was approved in the EU is Spherox by the company CO.DON. Spherox contains spheroids of human autologous chondrocytes for treatment of knee cartilage defects. The manufacturing process comprises the isolation of chondrocytes from a patient's cartilage biopsy by enzymatic digestion and expansion of cells in monolayer. After this, 3D spheroids of these chondrocytes are generated which are then the active substance of the product [37, 38]. This complex process, that contains the isolation of chondrocytes from the tissue of the patient, in vitro expansion, preparation of device for injection, shipping back to the clinic and aseptic application by a surgeon, is only possible because the entire procedure is highly automated.

However, future developments in other fields may help to improve safety and efficacy of cultivation processes and lead to more approved Tissue Engineering applications.

Questions
13. What are the two main applications for 3D cell culture?
14. What cultivation approach is the most widely used in 3D in vitro models?
15. How is a Tissue Engineering process structured?
16. What is the so-called 3 R rule and how is it connected to 3D cell culture?

1.6   Challenges of 3D Cell Culture

Although there is overwhelming evidence proving that cell culture on stiff 2D plasticware compared to three-dimensional models provides an inferior model regarding physiological cell behaviour and responses, it is still "gold standard" in many laboratories.

One of the most obvious reasons is simply the increased time and costs that arise with many 3D cultures. Mass produced and therefore cheap cell culture reagents and equipment is made for 2D culture and in many cases cannot be used for three-dimensional cultures. Many three-dimensional set-ups also need dynamic culture, to increase nutrient supply and waste removal, which is implemented with bioreactors, pumping systems or special incubation systems. Also, the preparation and handling of 3D cultures are more time-consuming. Cells seeded in a culture dish attach within minutes to a few hours, whereas matrix and matrix-free 3D cultures can take up days to be mature for further experiments [39]. Matrix or mould preparations like sterilization and cleaning add additional time to the workflow.

The lack of standardized protocols makes it also difficult to have comparable research, all over the scientific community. With mass produced and standardized tissue culture plastics, well-defined medium and cell culture SOPs as currently used for 2D cultures, it is rather easy to compare two experiments, even from across the globe. Thus, it is of great importance to establish similar resources of comparable matrices and handling procedures as for traditional 2D cell culture.

However, one of the most crucial aspects is the analysis of 3D grafts, as it poses rather big challenges. Light microscopy and fluorescence-based assays are essential parts of cell culture and can be easily applied to monitor cell monolayers. In 3D, however, major limitations arise due to the nature of multi-layered constructs, starting already with the graft exceeding the working distance of the microscope. Often only the outermost cell-layer can be observed leading to a biased conclusion. Furthermore, opaqueness, light scattering and fluorescent backgrounds all make imaging unreliable [40]. Advanced optical analysis with a confocal microscope can overcome these problems by imaging layer by layer, yet such expensive devices are not accessible for every research group. The same goes for analysis approaches adapted from tissue analysis, like histological sectioning and staining, micro-and-nano computer-tomography or magnetic resonance imaging.

Routinely used cell assays can be used but have to be adapted to work within the intended system. For example, widely used metabolic assays measure cell viability by intracellular conversion of special dyes. Although these assays like alamar blue$^{©}$ or MTT can be used for 3D cultures [41], it has to be considered that because of diffusion limitations cells from the core might not have access to the applied reagents. The same goes for testing methods that involve screening of soluble factors in the medium, secreted molecules might not reach the supernatant as efficiently as in a monolayer culture resulting in considerably lower values detected than actually secreted. Generally, the core will contribute less to the measured signal than the outer cell layers.

In the case of lytic assays like qPCR and Western Blot this problem is omitted, however it has to be ensured that complete lysis of all cells is achieved. Unfortunately, due to longer incubation times or overall harsher lysis methods that are needed to disassociate a 3D cell construct, sensitive analytes like proteins or RNA can be lost.

A more detailed description of challenges associated with analysis and approaches to overcome them can be found in Chap. 10.

If we consider 3D culture not only in a research setting with typically an end point analysis, but rather as means of producing cell-based therapeutics additional hurdles have to be considered. For potential application in real life scenarios like treatment of dystrophic epidermolysis bullosa or traumatic brain injury, extremely high cell numbers are needed (1–5 million per kg bodyweight [42, 43]). This again brings in the problem of time, cost and standardization, mentioned earlier. The more complex a cultivation procedure is, the more difficult it is to transform it into an automatable workflow. The culture systems have to be compatible with high-throughput screening in order to be used in a commercial, medical or pharmaceutical setting. Further problems arise during up-scaling. Cells can only be cultured into all three dimensions up to a certain degree. The gradient of nutrients and oxygen reaches zero at a certain distance away from the graft surface. Dynamic culture can improve the transfer of solutes to some degree, but this has also limitations. In native tissue this problem is overcome by the vasculature system. Nearly all tissues need to be vascularized in order to survive. Most cells cannot survive further away from a vessel than a few hundred micrometres [44]. This vascularization dilemma is one of the main obstacles in Tissue Engineering and a lot of work is done to first of all understand vessel formation in vivo and secondly to use the gained knowledge to implement it into an in vitro culture. This can be in the form of completely artificial vessels or neoformation of capillaries by heterocellular culture and addition of growth factors. More details on vascularization for 3D cell culture are found in Chap. 6.

Questions
17. What are economical and translational challenges of 3D cell culture?
18. What problems can occur in cell viability measurements of 3D cell constructs?

Take Home Message
- 2D cultivation does not represent the physiologic environment of cells.
- Results from 2D cultivation are often not predictable for the in vivo behaviour of cells in clinical trials.
- 3D cultures can be achieved using either matrix-free, matrix-based approaches or bioprinting.

(continued)

- The main application for 3D cell culture is the generation of relevant in vitro models and engineering of tissues for clinical tissue regeneration.
- All analyses usually do not work in 3D and thus must be optimized for 3D culture.
- Vascularization of 3D cultures is a major challenge in the generation of large-scale constructs.

Answers
1. 3D cell culture enables researcher to develop more relevant in vitro models.
2. 3D cell culture better resembles the in vivo situation. Thus, the data derived from 3D cell culture is more relevant than data derived from 2D cell culture
3. The development of novel therapies often needs to be stopped because the therapeutic effects observed in pre-clinical testing are not observed anymore in animal or human trials.
4. The most abundant one is collagen, which can resist tension, thus helping maintain structural strength. Other fibrous proteins are elastin which is important for tissue elasticity, fibronectin as linker between cells and other ECM molecules and laminin that is important for epithelial integrity by forming sheet-like structures in the basal lamina.
5. We can distinguish between physical (partial gas pressure, temperature, viscosity), chemical (nutrients, growth factors, salts/ions, toxins), mechanical (compression, tension, shear) and biological stimuli (cell-to-cell contacts, cell-to-ECM binding).
6. Cells can connect to other cells by forming tight junctions (barrier); they can form desmosomes and adherens junctions (mechanical stability) and gap junctions (communication).
7. The cells need a substrate to adhere to, medium which keeps a constant pH in combination with supplied CO_2, and provide them with basic salts, amino acids, glucose and vitamins (often supplemented with serum), a humid and warm (37 °C) environment provided by incubators and regular sub culturing.
8. Cells in 2D have a forced apical basal polarity as they are only able to adhere to the substrate below them, nearly unlimited space causes high substrate interaction and only in a very small lateral region of the cell are contacts to other cells possible. In 3D there is except for certain specialized tissues no polarization of the cell, a high number of cell-to-cell contacts is established, and the cells can self-organize into 3D structures.

(continued)

9. Matrix-free (spheroids, aggregates), matrix-based (cells plus scaffold) and bioprinting.
10. The most widely methods for the generation of aggregates are hanging drop, ultra-low attachment plates, microcavity plates.
11. Matrix-free cultivation is easy in handling (many commercial systems available), enables for high throughput.
12. For almost every application there is a suitable scaffold. The size and shape of the scaffold can be adjusted and larger constructs than with spheroids are possible.
13. In vitro models and Tissue Engineering.
14. Spheroid cultivation.
15. Cells are collected from the patient, expanded, seeded on a scaffold, and optionally cultivated in a bioreactor to achieve tissue maturation. Then, the tissue can be used as an in vitro model or as a transplant.
16. 3 R rule is a guideline concerning animal testing and stands for "replace, reduce and refine," 3D models have the potential to replace or at least reduce the number of animal testing.
17. Higher costs (custom made equipment, longer cultivation times, time intensive), less experience in the scientific community, complex protocols difficult to scale up and automatize.
18. Lower signals detected due to diffusion limitations of reagent or dye.

References

1. Ronfard V, Vertès AA, May MH, Dupraz A, van Dyke ME, Bayon Y. Evaluating the past, present, and future of regenerative medicine: a global view. Tissue Eng Part B Rev. 2017;23 (2):199–210. https://doi.org/10.1089/ten.teb.2016.0291.
2. Ledford H. Translational research: 4 ways to fix the clinical trial. Nature. 2011;477(7366):526–8. https://doi.org/10.1038/477526a.
3. Fernandez-Moure JS. Lost in translation: the gap in scientific advancements and clinical application. Front Bioeng Biotechnol. 2016;4:43. https://doi.org/10.3389/fbioe.2016.00043.
4. Justice BA, Badr NA, Felder RA. 3D cell culture opens new dimensions in cell-based assays. Drug Discov Today. 2009;14(1–2):102–7. https://doi.org/10.1016/j.drudis.2008.11.006.
5. Thoma CR, Zimmermann M, Agarkova I, Kelm JM, Krek W. 3D cell culture systems modeling tumor growth determinants in cancer target discovery. Adv Drug Deliv Rev. 2014;69–70:29–41. https://doi.org/10.1016/j.addr.2014.03.001.
6. O'Brien FJ. Biomaterials & scaffolds for tissue engineering. Mater Today. 2011;14(3):88–95. https://doi.org/10.1016/S1369-7021(11)70058-X.
7. Hynes RO. Cell adhesion: old and new questions. Trends Genet. 1999;15(12):M33–7. https://doi.org/10.1016/S0168-9525(99)01891-0.
8. Alberts B, et al. Molecular biology of the cell, vol. 20. 6th ed. Boca Raton: Garland Science; 2014.

9. Ricard-Blum S. The collagen family. Cold Spring Harb Perspect Biol. 2011;3(1):4978. https://doi.org/10.1101/cshperspect.a004978.

10. Mithieux SM, Weiss AS. Elastin. Adv Protein Chem. 2005;70:437–61.

11. Adams JC, Watt FM. Regulation of development and differentiation by the extracellular matrix. Development. 1993;117(4):1183–98.

12. Durbeej M. Laminins. Cell Tissue Res. 2010;339(1):259–68. https://doi.org/10.1007/s00441-009-0838-2.

13. McKeown SR. Defining normoxia, physoxia and hypoxia in tumours—implications for treatment response. Br J Radiol. 2014;87(1035):20130676. https://doi.org/10.1259/bjr.20130676.

14. Khorshid F. The effect of the medium viscosity on the cells morphology in reaction of cells to topography-I, pp. 15–17; 2004.

15. Jedrzejczak-Silicka M. History of cell culture. In: New insights into cell culture technology. London: InTech; 2017.

16. Vincent JFV. Structural biomaterials. New York: Macmillan; 1982.

17. Lodish H, Berk A, Zipursky SL, Matsudaira P, Baltimore D, Darnell J. Molecular cell biology. 4th ed. New York: W.H.Freeman; 2000.

18. Goodenough DA, Paul DL. Gap junctions. Cold Spring Harb Perspect Biol. 2009;1(1):002576. https://doi.org/10.1101/cshperspect.a002576.

19. Rodríguez-Hernández CO, et al. Cell culture: history, development and prospects. Int J Curr Res Acad Rev. 2014;2(12):188–200.

20. Kasper C, Charwat V, Lavrentieva A. Cell culture technology. Cham: Springer; 2018.

21. Amstein CF, Hartman PA. Adaptation of plastic surfaces for tissue culture by glow discharge. J Clin Microbiol. 1975;2(1):46–54.

22. Ladoux B, et al. Strength dependence of cadherin-mediated adhesions. Biophys J. 2010;98 (4):534–42. https://doi.org/10.1016/j.bpj.2009.10.044.

23. Evans N, et al. Substrate stiffness affects early differentiation events in embryonic stem cells. Eur Cell Mater. 2009;18:1–14. https://doi.org/10.22203/eCM.v018a01.

24. Zustiak S, Nossal R, Sackett DL. Multiwell stiffness assay for the study of cell responsiveness to cytotoxic drugs. Biotechnol Bioeng. 2014;111(2):396–403. https://doi.org/10.1002/bit.25097.

25. Discher DE. Tissue cells feel and respond to the stiffness of their substrate. Science. 2005;310 (5751):1139–43. https://doi.org/10.1126/science.1116995.

26. Baker BM, Chen CS. Deconstructing the third dimension – how 3D culture microenvironments alter cellular cues. J Cell Sci. 2012;125(13):3015–24. https://doi.org/10.1242/jcs.079509.

27. Mao AS, Shin J-W, Mooney DJ. Effects of substrate stiffness and cell-cell contact on mesenchymal stem cell differentiation. Biomaterials. 2016;98:184–91. https://doi.org/10.1016/j.biomaterials.2016.05.004.

28. Meyers J, Craig J, Odde DJ. Potential for control of signaling pathways via cell size and shape. Curr Biol. 2006;16(17):1685–93. https://doi.org/10.1016/j.cub.2006.07.056.

29. Ishii I. Histological and functional analysis of vascular smooth muscle cells in a novel culture system with honeycomb-like structure. Atherosclerosis. 2001;158(2):377–84. https://doi.org/10.1016/S0021-9150(01)00461-0.

30. McBeath R, Pirone DM, Nelson CM, Bhadriraju K, Chen CS. Cell shape, cytoskeletal tension, and RhoA regulate stem cell lineage commitment. Dev Cell. 2004;6(4):483–95. https://doi.org/10.1016/S1534-5807(04)00075-9.

31. Carrel A. A method for the physiological study of tissues in vitro. J Exp Med. 1923;38(4):407–18. https://doi.org/10.1084/jem.38.4.407.

32. Keshaw H, Forbes A, Day RM. Release of angiogenic growth factors from cells encapsulated in alginate beads with bioactive glass. Biomaterials. 2005;26(19):4171–9. https://doi.org/10.1016/j.biomaterials.2004.10.021.

33. Breslin S, O'Driscoll L. Three-dimensional cell culture: the missing link in drug discovery. Drug Discov Today. 2013;18(5–6):240–9. https://doi.org/10.1016/j.drudis.2012.10.003.

34. Elliott NT, Yuan F. A review of three-dimensional in vitro tissue models for drug discovery and transport studies. J Pharm Sci. 2011;100(1):59–74. https://doi.org/10.1002/jps.22257.

35. Ronaldson-Bouchard K, Vunjak-Novakovic G. Organs-on-a-chip: a fast track for engineered human tissues in drug development. Cell Stem Cell. 2018;22(3):310–24. https://doi.org/10.1016/j.stem.2018.02.011.

36. Skardal A, Shupe T, Atala A. Organoid-on-a-chip and body-on-a-chip systems for drug screening and disease modeling. Drug Discov Today. 2016;21(9):1399–411. https://doi.org/10.1016/j.drudis.2016.07.003.

37. Bartz C, Meixner M, Giesemann P, Roël G, Bulwin G-C, Smink JJ. An ex vivo human cartilage repair model to evaluate the potency of a cartilage cell transplant. J Transl Med. 2016;14(1):317. https://doi.org/10.1186/s12967-016-1065-8.

38. Eschen C, et al. Clinical outcome is significantly better with spheroid-based autologous chondrocyte implantation manufactured with more stringent cell culture criteria. Osteoarthr Cartil Open. 2020;2(1):100033. https://doi.org/10.1016/j.ocarto.2020.100033.

39. Kapałczyńska M, et al. 2D and 3D cell cultures – a comparison of different types of cancer cell cultures. Arch Med Sci. 2018;14(4):910–9. https://doi.org/10.5114/aoms.2016.63743.

40. Costa EC, Silva DN, Moreira AF, Correia IJ. Optical clearing methods: an overview of the techniques used for the imaging of 3D spheroids. Biotechnol Bioeng. 2019;116(10):2742–63. https://doi.org/10.1002/bit.27105.

41. Bonnier F, et al. Cell viability assessment using the Alamar blue assay: a comparison of 2D and 3D cell culture models. Toxicol In Vitro. 2015;29(1):124–31. https://doi.org/10.1016/j.tiv.2014.09.014.

42. Hook K. Stem cells and respiratory disease. Lancet Respir Med. 2017;5(3):178–9. https://doi.org/10.1016/S2213-2600(17)30056-5.

43. Rashidghamat E, et al. Phase I/II open-label trial of intravenous allogeneic mesenchymal stromal cell therapy in adults with recessive dystrophic epidermolysis bullosa. J Am Acad Dermatol. 2019;83:447. https://doi.org/10.1016/j.jaad.2019.11.038.

44. Frerich B, Lindemann N, Kurtz-Hoffmann J, Oertel K. In vitro model of a vascular stroma for the engineering of vascularized tissues. Int J Oral Maxillofac Surg. 2001;30(5):414–20. https://doi.org/10.1054/ijom.2001.0130.

Lab Equipment for 3D Cell Culture

2

Sebastian Kreß, Ciarra Almeria, and Cornelia Kasper

Contents

S. Kreß · C. Almeria
University of Natural Resources and Life Sciences, Vienna, Austria
e-mail: Sebastian.kress@boku.ac.at; ciarra.almeria@boku.ac.at

C. Kasper (✉)
Institute of Cell and Tissue Culture Technologies, University of Natural Resources and Life Sciences, Vienna, Austria
e-mail: cornelia.kasper@boku.ac.at

© Springer Nature Switzerland AG 2021
C. Kasper et al. (eds.), *Basic Concepts on 3D Cell Culture*, Learning Materials in Biosciences,
https://doi.org/10.1007/978-3-030-66749-8_2

Abbreviations

CHAPS	3-[(3-cholamidopropyl)dimethylammonio]-1-propanesulfonate
CHO	Chinese hamster ovary
DEAE	Diethylaminoethyl
DNase	Deosxyribonuclease
E. coli	*Escherichia coli*
ECM	Extracellular matrix
ECS	Extracapillary space
EDTA	Ethylenediaminetetraacetic acid
EtO	Ethylene oxide
HA	Hyaluronic acid
HF	Hollow fibers
HHP	High hydrostatic pressure
hiPSC	human induced-pluripotent stem cells
ICS	Intracapillary space
MSC	Mesenchymal stem cell
MWCO	Molecular weight cut-off
PAA	Poly(acrylic acid)
PCL	ε-poly(caprolactone)
PEG	Polyethylene glycol
PEO	Poly(ethylene oxide)
PIPPAAm	Poly-N(isopropylacrylamide)
PLA	Poly(lactic acid)
PLGA	Poly(lactic-co-glycolic) acid
PVA	Poly(vinyl alcohol)
RNasE	Ribonuclease
SDS	Sodium dodecyl sulfate
SDS	Sodium deoxycholate
STR	Stirred tank reactor
TE	Tissue engineering
TIPS	Thermally induced phase separation
UV	Ultraviolett (light)
VEGF	Vascular endothelial growth factor

What You Will Learn in This Chapter
When starting to work with 3D cell cultures it is essential to be aware of the differences to common 2D cultures in terms of expenses and equipment as well as the type and dimension of 3D you intend to work with. The various culture methods require different equipment, efforts, and awareness on handling and analytics. Therefore, it is crucial to choose the right structure and architecture, the respective carrier/scaffold, the equipment for generating and culturing the models, and the methodology to manipulate and analyze them to be able to perform the process for the application answering the hypothesis. There is a wide range of commercially available lab equipment, bioreactors, consumables, carriers/scaffolds, and assays necessary to establish, perform, and analyze 3D cell culture. Handling, manipulation, and analysis of 3D cultures, differing significantly from 2D cultures, with the respective requirements and efforts will be considered as well. An overview on scaffold free 3D cell culture will be provided in Chap. 9.

2.1 Considerations for Working in 3D

Previously in Chap. 1.3, the main differences in 2D and 3D culture techniques and the relevance to bid farewell to classic 2D culture have been presented. Traditional 2D cell culture models provide highly artificial conditions and do not only harm the cells but also alter gene expression, cell behavior, and thereby also their response to assays performed on them. This is especially critical for applied clinical research and development of therapeutic strategies as the results are compromised and thus less relevant. Pharmaceutical research already suffers from false positive and false negative results as 2D cultures allow high throughput screenings but lack physiological relevance. Thereby, the predictability of the outcome applying the same substances tested in 2D in animal studies is poor, limiting the transferability of the results gained. Thus, rendering 2D not obsolete but only as a weak indication.

In contrast, in 3D models using tissue-specific extracellular matrix (ECM) components and architecture, cells exhibit biochemical and morphological features specific for the in vivo state that are not or differently expressed in 2D. Furthermore, time and spatially resolved mechanical and biochemical signals can be introduced due to the arrangement of the cells. Thus, the big impact of 3D culture is the possibility to culture cells in a more physiological environment allowing the observation of natural cellular behavior, increasing the relevance and transferability of the results onto the in vivo situation. However, 3D culture is complex, expensive, and labor-intensive. There are more and more commercially available scaffolds but most applications demand custom-made fabrications mimicking the tissue-specific requirements. Therefore, specific equipment, systems, and machines need to be acquired for the fabrication of scaffolds with organ or tissue-specific characteristics from various materials. Equipment, material, and fabrication techniques to manufacture various scaffolds as well as equipment and bioreactor systems for 3D culture are discussed within

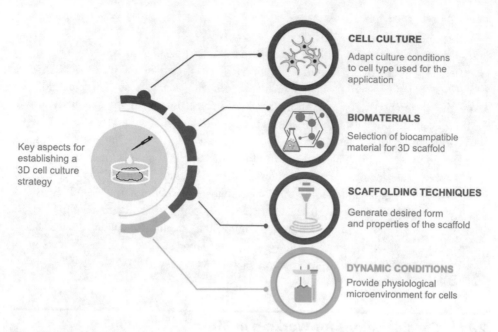

Fig. 2.1 3D Cell culture comprises four major key aspects. The combination of cells, biomaterials, and scaffolding as well as dynamic culture conditions has to be in accordance to achieve physiological conditions in vitro mimicking the native microenvironment of cells. Each aspect is defined by distinct challenges and needs to be optimized and adapted to the specific application

this chapter as well as different strategies to establish 3D cultures (as depicted in graphical abstract in Fig. 2.1).

2.2   Biomaterials for 3D Cell Cultivation

Biomaterials have served as a crucial utility in the field of regenerative medicine in order to promote cellular growth and differentiation as well as healing of damaged tissues, which has further evolved in the promising field of tissue engineering (TE). Principally, biomimetic materials (3D scaffolds) should offer a structure for cell adhesion, proliferation, and ECM formation until restoration of necrotic tissue is achieved. According to the Consensus Development Conference in the 1980s (Chester, UK, 1982), biomaterials were defined as "any substance, other than a drug, or a combination of substances, synthetic or natural in origin, which can be used for any period of time, as a whole or as a part of a system, which treats, augments or replaces any tissue, organ or function of the body" [1, 2].

One pre-requisite for 3D in vitro models are carrier structures to best mimic the physiological environment of the cell's origin (see Chap. 1.2) and support the natural behavior of the cells. Scaffolds for TE must possess adequate mechanical integrity including structural (macro- and microstructural properties) and mechanical characteristics (mechanical strength, elasticity, and stiffness) similar to that of the target tissue and support the physiological load of the body until remodelling process is completed. Furthermore, its mechanical properties shall also be sufficient enough to withstand all necessary surgical manipulations during implantation. All these properties have strong impact on cell survival, adhesion, proliferation, differentiation as well as vascularization and gene regulation and are needed to be adjusted according to the target tissue and specific application [3].

Biomaterials can be synthetic or of natural origin and should be characterized by high biocompatibility, known degradation kinetics and by-products, material biomimicry and proper structural and mechanical attributes. Furthermore, natural ECM materials present ligands for cellular adhesion, whereas synthetic materials lack these chemical groups facilitating cell-ECM attachment. In the latter case, cellular binding sequences must be incorporated by, e.g., protein adsorption. Moreover, cell adherence and motility are also defined by ligand density within a scaffold and the pore size. Beyond adherence and motility, the surrounding ECM influences the cellular behavior by transmission of stimuli from mechanical, structural, and compositional cues. Exemplary mesenchymal stem cell (MSC) differentiation can be driven towards the myogenic lineage by culturing on an elastic gel mimicking muscle or osteogenic culturing on a rigid material [4]. Materials provide different intrinsic physicochemical properties such as molecular weight, chemical composition, hydrophilicity, mechanical properties, surface chemistry, and degradation rate in addition to biochemical properties such as cytocompatibility, enabling cellular attachment, motility as well as proliferation without impeding the intended cellular function or eliciting immune responses. A more in-depth description of materials and their distinctive characteristics and fields of application is described in Chap. 4. Moreover, increasing focus has been directed towards the fabrication of composite scaffolds, consisting of two or more materials that present different properties. Hence, generation of scaffolds, that indicate distinct characteristics from their original materials, are enabled.

The material's characteristics are further refined during scaffolding. The composition, orientation, and architecture of the materials utilized add a new dimension to their intrinsic characteristics adjusting the scaffold properties in terms of, e.g., porosity, elasticity, rigidity, plasticity, ductility but also cytocompatibility and cellular motility as well as functionality. Additionally, not only the various materials have different properties but also the distinct scaffolds exhibit specific advantages and disadvantages. The choice on material and scaffolding technique affects/excludes each other depending on one or the other but is primarily defined by the application and the scaffold properties that have to be exhibited for the specific cell culture application. The selection process on material and scaffolding is top-down, initially becoming apparent on the application and cell/tissue model to serve best for assays and analysis to proof or de-bunk the stated hypothesis. Thereafter, the suitable scaffold properties are defined to house the necessary cells assuring their functionality that might be based on their spatial arrangement.

2.3   Concepts of Cell Culture Techniques for TE

Researches in the fields of TE and regenerative medicine have shifted from 2D models to 3D strategies as described in Chap. 1. Therefore, biomaterials have gained considerable relevance in the success of tissue replacement or regeneration as they demonstrate to have the ability to influence biological processes which are necessary for tissue regeneration. 3D scaffolds can be manufactured from different materials and can be shaped in various forms. The main challenge remains the choice of suitable materials that demonstrate appropriate characteristics, as mentioned above, and could deliver satisfactory performance, depending on the tissue of interest. Nowadays, several types of biomimetic materials such as ceramics, composites, and natural or synthetic polymers are currently used for in-depth studies for different TE strategies [2, 5, 6]. Therefore, the different types of scaffolds that are abundantly investigated and used for TE applications (see Table 2.1) up to date are introduced on the following paragraphs as well as commercially available systems are presented.

2.3.1   2 ½ -Dimensional Cell Culture Strategies.

Cell-sheet layering has been employed as a tool for regenerative medicine for heart, cornea, and cartilage treatments [12, 13]. In traditional TE, cells are seeded onto 3D scaffolds composed out of polyglycolic acid, gelatin, or collagen which are mostly biodegradable. However, 3D scaffolds demonstrate a few limitations including inhomogenous cell distribution on the matrix, induction of inflammatory responses, occurrence of fibrosis arising

Table 2.1 Exemplary selection of relevant biomaterials for 3D cell culture application

Biomaterial	Applications	Commercial availability	References
Collagen fleece	TE, ocular surgery, skin and orthopedics regeneration, drug delivery systems, wound dressing	Matriderm® (Dr. Otto Suwelack Skin & Health Care AG, Germany)	[7]
Bioactive glass	Hard tissue replacement, orthodontic application	MedPor®PLUS™ (Porex surgical, USA), PerioGlas®	[8]
Hydrogel	Connective, soft and hard TE, gene therapy, drug delivery implants, orthopedic regeneration	Matrigel (thermo fisher scientific, USA) PLMatrix (PL bioscience GmbH, Germany) GelMa (CellInk, USA)	[9]
Silk sponge or foam	Orthopedics TE, wound dressing, cardiac repair	Silk fibroin solution (advanced BioMatrix, USA)	[10]
Polymer fibers	Cardiovascular TE, wound healing, high-throughput cell culture	NanoECM™ (Nanofiber solutions, Inc., USA)	[11]

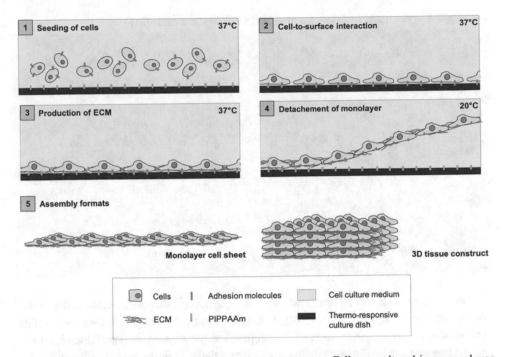

Fig. 2.2 Schematic image of the cell sheet engineering process. Cells are cultured in a monolayer until confluency is reached (>80%). Cells have fully connected to each other via cell-to-cell junction proteins by then and deposited the extracellular matrix (ECM). With the use of temperature-responsive culture dishes covalently bonded with polymer, namely poly-N(isopropylacrylamide) (PIPPAAm), confluent cells are released as cell sheets at temperatures <20 °C, preserving formed cell-to-cell connections and deposited ECM. With this, disruption of cell membrane proteins and ECM by enzymatic treatment is avoided. Harvested cell sheets can be stacked together to create a multilayer 3D constructs and enable reconstruction of native tissues

from the degradation of scaffolds and production of cell-sparse tissue. This aspect is especially important for regenerative processes which are highly dependent on cell–cell interactions such as for cardiac TE. To circumvent these limitations cell sheet layering has evolved as a promising approach for 3D cardiac tissues by layering cardiac monolayer cell sheets and has shown improvements towards cardiac function of a diseased heart [14]. However, to establish a 3D tissue by cell sheet layering, the method still mostly relies on 2D culture. The principle is to culture cells on a matrix in a 2D fashion and after maturation of the cellular layer several layers are stacked above each other to establish a 3D tissue. In detail, this technique uses cells seeded on a cell surface which is covalently bonded with a temperature-responsive polymer, namely poly-N (isopropylacrylamide) (PIPPAAm) (see Fig. 2.2). Cells are cultivated until they reach confluency and by lowering the temperature from 37 °C to 32 °C the surface turns from a slightly hydrophobic into a highly hydrophilic state. Ultimately, this results in rapid swelling of the PIPPAAm, forming a hydration layer between the surface and the expanded cells, and leads to the detachment of the cells from the PIPPAAm-coated surface. In comparison with proteolytic

Table 2.2 Commercially available microcarrier of different materials [21]

Name	Company	Size (μm)	Density (g/L)	Material
Cytodex-1 [22, 23]	GE Healthcare	60–87	1.03	Dextran matrix with positively charged diethylaminoethyl (DEAE) groups
Cytophore 1	Pharmacia	200–280	1.03	Crosslinked cotton cellulose with diethylaminoethyl (DEAE) groups
CultiSpher G [24]	Sigma-Aldrich	130–380	1.04	Crosslinked porcine gelatin matrix
TSKgel Tresyl-5PW	Tosoh bioscience	65 ± 25	1	Hydroxylated methacrylate matrix with Tresyl ligand derivatized with protamine sulfate (primary amine)
Hillex [25]	Pall SoloHill	150–210	1.1	Dextran matrix with treated surface
ProNectin	Pall SoloHill	150–210	1.03	Polystyrene matrix

cell harvesting techniques by enzymatic digestion, intact cell sheets can easily be harvested non-invasively along with their deposited ECM and without damaging critical cell surface proteins such as ion channels and cell-to-cell junction proteins. Moreover, these cell sheets are stacked and can be directly applied to the target tissue to promote cell–cell contact and adhesion efficiently, as deposited ECM is still present underneath the multi-layered cell sheet. Nevertheless, as promising as this method seems for TE applications, it reveals drawbacks in regard to stability of the cell sheets and demonstrates labor-intensive work for production (2.5–5 h fabrication duration of a five-layered cell sheet) [15, 16].

Microcarriers provide supportive structures for cells and are composed of natural (gelatin, collagen, and cellulose) or synthetic (plastic, glass, or dextran) materials (list of commercially available microcarriers in Table 2.2). Cell carriers provide a relatively high surface to volume ratio facilitating the culture of high cell densities in low volume similar to aggregate culture but providing a structure for cell adherence and growth. However, these mostly provide a spherical 2D surface for cell attachment to facilitate suspension culture. Therefore, most cell carrier approaches might be considered 2 ½ D as they go beyond the classical static 2D plastic surface culture but do not provide a real 3D environment. For porous microcarriers, their microstructural properties such as porosity can be engineered as interconnective pores in a spherical architecture. Herein, cells are allowed to attach and maximize their proliferative potential on the surfaces and inner pores. The surface area of 1 g of microcarriers is equivalent to the surface of fifteen 75 cm^2 cell culture-treated flasks. Firstly, microcarriers were developed for the cultivation of anchorage-dependent mammalian cells in order to promote cellular growth for high cell densities. Soon after, they were made suitable for the growth of almost all cell lines by introduction of surface modifications with poly-Lysine, gelatin, or ProNectin to enhance cell adhesion and growth. Additionally, microcarriers can be coated with other chemical factors to improve cell adhesion including fibronectin or factors contained in culture medium formulations. This type of cultivation strategy facilitates a more expansive monolayer environment for the cells. Large quantities of microcarrier covered with cells can be suspended in dynamic, environmentally controlled cultivation systems such as stirred tank bioreactors and make nutrient and gas transfer more efficient compared to 2D culture. Nevertheless, microcarriers can only be considered a 2 ½ D model as cells still grow on a surface and do not mimic the natural environment of the cells. Furthermore, the low area-to-volume ratio compromises the potential of maximum cell loading per sphere, aggregations of cell-loaded microcarriers (as seen in Fig. 2.3f) and cells on the surface are more vulnerable to bead collision or shear forces in the culture. Cultivation parameters including agitation rate in such dynamic systems need to be controlled as cell damage can occur easily. Another limitation demonstrates the reduced nutrient and oxygen mass transfer in the inner core of porous spheres and therefore the fitness and quality of the cultivated cells. It has also been reported that the surface coating and culture medium used in microcarrier-based cultures depend on the type of cells to be expanded, due to the different specific adhesion molecules expressed by the cells. Therefore, the selection of microcarrier materials and coatings needs to be adapted to the cell type to be used [17–19].

2.3.2   Soft 3D Scaffolds

Hydrogels are highly hydrated polymers (water content >25%) and demonstrate ECM-like properties which facilitate efficient cell encapsulation. Furthermore, hydrogels are typically biodegradable and provide an environment similar to that of the native ECM of many tissues. They are used as delivery vehicles for bioactive agents and 3D structures for homing of cells or tissue replacements and provide stimuli for tissue regenerative processes.

They are classified into two groups according to their hydrophobic polymer chains: a) naturally derived hydrogels including collagen, agarose, fibrin, gelatin, chitosan, alginate, cellulose, and hyaluronic acid (HA) and b) synthetically derived hydrogels such as polyethylene glycol (PEG), poly(ethylene oxide) (PEO), poly(vinyl alcohol) (PVA), poly (acrylic acid) (PAA), and polypeptides. The hydrophilic and mechanically stable characteristics of hydrogels permit cells to be in an aqueous environment without the risk of dissolving. Furthermore, efficient exchange of nutrient, oxygen, waste, and other water-soluble molecules is given due to their high permeability. Natural hydrogels preserve the macro-molecular properties of natural ECM, which make them an attractive biomimetic material. On the other hand, synthetically derived hydrogels indicate appealing advantages including their high reproducibility and controllability of specific chemical and mechanical properties such as degradable linkages, molecular weight and crosslinking modes. Manipulation of the hydrogel properties determines the polymerization dynamics, mechanical integrity, crosslinking density, and degradation properties. However, both groups of hydrogels reveal limitations as a tool for TE. Natural hydrogels are highly variable due

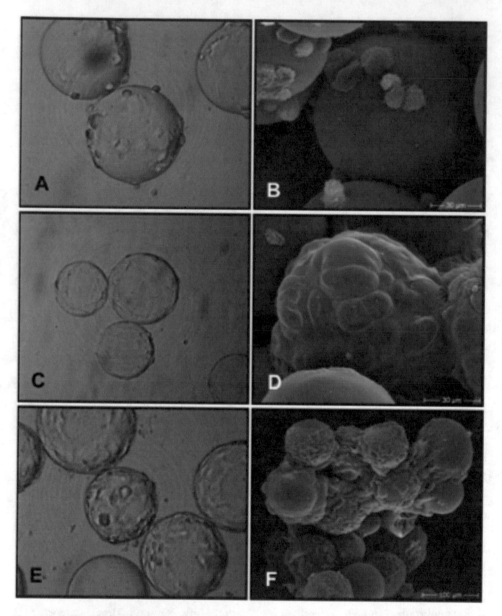

Fig. 2.3 Cell-loaded Cytodex type 1 microcarrier during spinner flask cultivation. Images depict microcarrier culture taken by light microscopy (**a**), (**c**), (**e**) (magnification 100×) and SEM (**b**), (**d**), (**f**) at different time points: day 0 (**a**), (**b**), day 14 (**c**), (**d**), and day 28 (**e**, **f**) [20]

to their native source and fabrication parameters designed in specific research groups, which makes comparison between different studies difficult. Whereas, synthetic hydrogels lack major components such as cell adhesion and migration molecules, which are normally

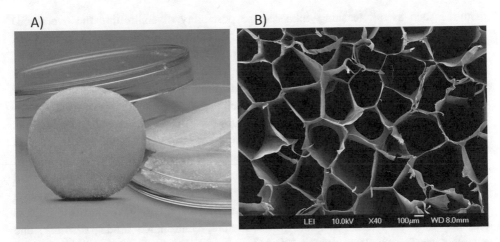

Fig. 2.4 Collagen sponge for TE applications. (**a**) Collagen sponge SpongeCol® (Advanced BioMatrix, Carlsbad, CA, USA) (4–21 mm discs and 1.5 mm thickness) fits into 96-well culture plate well and is sterilized by irradiation. Furthermore, they are crosslinked to increase mechanical stability and durability. (**b**) Porous network of SpongeCol® allows cells, nutrients, and waste products to diffuse sufficiently while increasing surface area for cell attachment, growth, and migration. Pore sizes range between 100 and 400µm, with an average size of 200µm. Scale bar: 100µm

provided by the native ECM for necessary cell–cell junctions. Selection of an appropriate material and scaffold design needs to be adapted to the specific application in order to address the biological variables. A more detailed description of hydrogels for TE applications is provided in Chap. 5 [26–28].

Sponge or foam scaffolds are porous constructs (Fig. 2.4), which are fabricated via controlled freezing–drying (see Chap. 2.4.5) and solvent casting/particulate leaching of various concentrations (see Chap. 2.4.3) of the chosen biomaterial solutions such as chitosan. Additionally, gelatin, for example, is hydrolyzed collagen and crosslinked with proteins for stabilization of structural integrity and mechanical strength. The interconnected pore structure and pore size can be controlled or randomly formed depending on the fabrication method applied. Sponge scaffolds have found applications in several fields including repair of nasal malformations, bone formation, cartilage development, and ligament replacement as well as for joint pain, inflammation, diabetes, heart disease, and wound dressings. It is also used as a tool for controlled release systems to deliver growth factors onto the site of the target tissue as well as protect these factors from spontaneous proteolysis and allow a prolonged retention of activity in the tissue [29, 30].

Fibrous scaffolds are manufactured from various polymers including ε-poly (caprolactone) (PCL), poly(lactic acid) (PLA), poly(lactic-co-glycolic) acid (PLGA), gelatin, cellulose, or silk fibroins. Common fabrication methods of nanofibers are electrospinning, thermally induced phase separation (TIPS), or molecular self-assembly (see method description in Chap. 2.4). A combination of TIPS with other processing methods such as salt leaching enhances the possibility of controlling the overall 3D geometry and inner structure. They provide high porosity, surface-to-volume ratio and induce enhanced cellular adhesion and growth. Moreover, rapid diffusion of (optional) incorporated bioactive molecules and cell infiltration is facilitated, similar to sponge or foam scaffolds as aforementioned. Functionalization of fiber surfaces with different ligands, including proteins, ceramics, and proteins is also commonly performed as synthetic biomaterials lack specific functional groups for cellular attachment and migration. Nevertheless, nanofibers demonstrate potent templates for promoting the development of in vivo-like phenotypes of cells as they mimic the native structure and mechanical properties of the ECM. Various cell types including chondrocytes, osteoblasts, and hepatocytes demonstrated enhanced cellular attachment, proliferation and differentiation on nanofibrous materials compared to 2D platforms. Lastly, they have shown to possess biocompatible properties, adequate microstructure, controllable biodegradability, and excellent mechanical characteristics [29, 31].

2.3.3 Hard 3D Scaffolds

Ceramic biomaterials are usually composed of inorganic calcium or phosphate salts and produced by salt leaching, TIPS, gel casting, or 3D printing (see method description in Chap. 2.4). They are classified into three subtypes: (a) inert (non-absorbable), (b) semi-inert (bioactive), and (c) non-inert (resorbable). They successfully stimulate cellular growth and formation of bonds between the cells and target tissue. The most advantageous characteristics are their osteo-inductive properties to which they are widely used in bone TE and dental implant applications. Moreover, ceramic scaffolds have shown great biocompatibility and bio-resorbability in vitro as well as in vivo. Nonetheless, the brittleness of ceramics and lower mechanical strength compared to that of human bone (100–230 MPa) at high degrees of porosity limit the effectiveness of such templates for bone regeneration. Some research groups tried to crosslink a second-phase addition of different materials including polymer fibers, particles (e.g., nanocrystals), and whisker (a filament of material with high tensile strength up to 10–20 GPa) or develop hierarchically porous structure to improve the interior structural integrity and toughness. Furthermore, other ceramics such as calcium sulfate (CSH) demonstrate a high degradation rate compared to the formation of new bone. As a consequence, acidic by-products are released into the tissue which negatively affect cellular proliferation and viability. However, it was suggested that the pace of new bone regeneration could be matched with that of the biomaterial's degradation rate by controlling the size of crystalline grain, a vital component to improve bone graft healing, in the construct. The diverse range of possible chemical compositions and controllable structural properties of ceramic scaffolds indicate superior advantages for a biomimetic material [29, 32].

Bioactive glasses (BG) have gained great attention especially for bone TE applications. The first composition in this biomaterial class, namely 45S5 glass, has been presented by

Larry Hench [33] 50 years ago. It enabled stimulation of osteogenesis via forming connections with the native bone and the release of bioactive ions from the 45S5 glass. Due to this success, many researchers were attracted to investigate and optimize the concept of BG for TE applications. Indeed, several implants have been subjected to millions of patients, globally since then. Furthermore, a large number of innovative compositions and other types of bioactive glass have been proposed to adapt to requirements of specific clinical application. Borate BG has attracted substantial interest besides the conventional silicate BG due to high dissolution rates, enhanced formation of calcium phosphate-based apatite on the surface, and high solubility in contact with biological fluids. Incorporation of other cations besides calcium and sodium is also highly encouraged to evoke beneficial effects including acceleration of self-repair kinetics and osteogenesis. No adverse or toxic effects have been observed with the addition of growth factors as they are easily excreted via body fluids and inorganic elements provide a cost-effective approach. For clinical applications, it was urged to confer to the requirements of surgeons which needed a format of BG that could be easily pressed into the bone defect. Therefore, 45S5 Bioglass® particulate was developed, made commercially available under the name of PerioGlas® (particle size range 90–710μm) (NovaBone Products LLC, Alachua, FL, USA). It was approved by FDA in 1993 and made available to the global market for jawbone-repair connected with periodontal diseases. However, the maximum potential of BG for biomedical applications is yet to be fully investigated and relevant markets are anticipated to continuously grow in upcoming years [8, 29].

Composite scaffolds combine various biomaterials such as materials mentioned in previous paragraphs (polymers/ceramics and synthetic/natural polymers, metals) in order to circumvent limitations that arose when using individual materials. They indicate to be highly relevant for biomedical engineering and TE applications. The efficacy of these composites has been reported by several studies since they present the required properties for TE and can be used for soft and hard tissue replacement and regeneration. Engineering desired mechanical and physiological characteristics including size, fraction, morphology, and arrangement of the reinforcing phase can be achieved in higher degrees along with the combination of different materials. For example, the degradation kinetic could be adapted as it has been the case for PCL. PCL normally exhibits slow degradation rates, weak mechanical strength, and poor cellular adhesion. However, by blending of PCL and other biomaterials such as cellulose, nanohydroxyapatite, carbon nanotubes, or cyclodextrins, these attributes were circumvented [29, 34].

2.3.4   Special Class of Biomaterial for TE

Decellularized natural tissues have become a favorable biomaterial used as 3D scaffold compared to synthetic scaffolds since the preferred chemical composition of biomaterials for TE should be comparable or as close to that of the host tissue as feasible. They enable interaction with the host cells/tissue, promote higher bioactivity and cellular recognition, leading to specific biological responses. Native tissues from the host consist of tissue-specific cells, growth factors, and ECM. The cells' deposited ECM is composed of various growth factors and cytokines including fibronectin, filaments, collagen type I and II, as well as proteoglycans, which evokes migration of cells (see Chap. 1.2) and facilitates mechanical signaling through receptors by providing an equivalent microenvironment to the native tissue. Collagen and proteoglycans contribute to most abundant proteins in the ECM and yield adequate tensile strength and durability of the tissue. Hereby, native cells of the ECM as well as all other cellular components are eliminated via a decellularization process, to avoid inflammatory responses to the host tissue, while other components and properties including structural integrity, biochemical assets, biological activity, composition, and hemocompatibility are preserved. Afterwards, desired cells are re-seeded (recellularization) or bioactive molecules incorporated into the decellularized ECM and implanted into the host to direct cell migration modulate cellular behavior and tissue-specific gene expression [35].

High availability of decellularized materials is present due to numerous possible sources including de novo ECM from autologous, allogenic, or xenogenic cells and native human or animal (bovine, porcine) tissues and organs. However, a few challenges arise with xenogenic and in some cases also allogenic-derived decellularized scaffolds as residues of its native components may appear after treatments and induce adverse effects in the host tissue. These aspects could put a patient into risk and might require immunosuppressant treatments in addition to existing health complications. Consequently, it is suggested that the biological activities are predetermined by the source and preparation methods of such matrices [36].

2.3.4.1 Decellularization Strategies

The decellularization process comprises different techniques in order to retain the original properties of the tissue or organ intact such as ECM, heart, lungs, urethra, and bladders [36, 37] (miromatrix, https://www.youtube.com/watch?v=UBGxvGAp878, https://www.youtube.com/channel/UCTnrPx3uGnUS9edzYFw3q-g). They can be differentiated into the following categories:

Chemical and Enzymatic

In principle, through this method cells and genetic material are solubilized by manipulating the cells' intrinsic charge and undesirable native components can be removed. Most common chemical agents are ionic surfactants such as sodium dodecyl sulfate (SDS), sodium deoxycholate (SD), or a non-denaturing detergent 3-[(3-cholamidopropyl) dimethylammonio]-1-propanesulfonate (CHAPS). Further chemicals for decellularization present acids and bases (peracetic acid and sodium hydroxide, respectively), ethylenediaminetetraacetic acid (EDTA), and enzymes such as trypsin, deoxyribonuclease (DNase), and ribonuclease (RNasE). After treatment with such agents, an extensive washing procedure needs to be performed in order to eliminate residual chemicals. In some cases, signaling proteins (e.g., GAGs and growth factors such as vascular endothelial growth factor (VEGF)) and the architecture of the tissue can also be damaged during the decellularization process, which could lead to incapability of the cells to adapt to the tissue and negatively affect the biochemical cues for the regulation of cellular function. It has been suggested to use a combination of chemical and enzymatic treatments to reduce the drawbacks of specific agents [38].

Physical and Mechanical

As certain concerns arise towards the toxic and destructive properties of chemical and enzymatic treatment, an alternative approach of physical and mechanical decellularization methods has been suggested. Generally, decellularization is achieved by eliminating the native tissue constituent cells and nuclear material via high hydrostatic pressure (HHP) (>600 MPa) and freeze-thawing (alternating temperatures between -80 °C to 37 °C). Nevertheless, similar limitations were reported for this approach due to the extensive washing step afterwards. Formation of ice crystals, denaturation of ECM proteins, alterations in protein content and structural integrity of the decellularized matrix were observed. In recent years, supercritical carbon dioxide (CO_2) (at 31.1 °C and 7.40 MPa) treatments were introduced for the decellularization process as CO_2 is non-polar and can easily diffuse from the tissue, removing extensive washing procedures required in the above-mentioned methods. Diffusion is accelerated by addition of ethanol in order to remove polar phospholipids of cell membranes. Moreover, all cellular debris and components were successfully removed while mechanical properties and protein content were preserved [37, 38].

2.3.5   Cell Seeding Strategies

With the generation of three-dimensional structures, plain superficial cellular seeding is not sufficient when a uniform cellular distribution throughout the whole scaffold is desired. For the various scaffolding techniques and resulting structures, different seeding strategies are presented.

At first, the scaffold has to exhibit cell adherence motifs and a surface charge facilitating cellular attachment. To facilitate or improve cellular adherence, biological glues, e.g. fibrin or fibronectin, mimicking extracellular matrix can be applied or heparin binding ligands or other cell-matrix signaling molecules can be incorporated.

The most straightforward approach is to seed cells on top of the scaffold facilitating migration inside. Therefore, the utilization of scaffolds needs the consideration of cell seeding techniques to spatially arrange the cells enabling the orchestration of their functionality.

Passive, static, or gravitational seeding, as used in 2D cell culture is simple and easy to perform, however, inefficient for 3D structures as superficially seeded cells only cover the scaffold's surface. In doing so, a cell suspension is applied onto the scaffold and incubated. Thereby, the cells attach to the surface of the scaffold regardless of the three-dimensionality. Intrusion of cells inside the scaffold is highly dependent on the material and porosity as well as active cell migration but will likely not be achieved by gravity as the cells tend to attach to the surface and then will have to migrate actively to also spread within the scaffold. In this case, mostly scaffold modification has to be carried out prior to seeding to orchestrate the cells' migration. Otherwise, the cells will reside mostly on the surface and only slowly migrate by themselves. However, this approach significantly prolongs the maturation of the tissue as the cells are not distributed throughout the scaffold properly, but they are required to self-organize. A semi-dynamic seeding strategy is the injection of cells inside the scaffold. Thereby, the distribution throughout the scaffold can be achieved more readily as the cells start migrating and spreading from within rather than from outside the scaffold. However, for this approach the scaffold architecture must allow the injection.

Dynamic seeding strategies increase efficiency in scaffold penetration.

Centrifugal seeding uses rotational systems creating centripetal forces to increase cellular scaffold infiltration. Thereby, a scaffold is rotated or spun with cells and culture medium. High speed rotational seeding increases penetration depth in less time but might affect cell morphology, in contrast to low speed. Another advantage of centrifugal seeding is the availability of centrifuges within a cell culture lab.

Magnetic cell seeding requires the attachment or incorporation of nanoparticles onto or into cells. Utilizing a magnet enables to attract the cells guiding them by electromagnetic forces. However, possible cytotoxicity of the utilized nanoparticles has to be excluded.

Another method is the utilization of pressure differential for seeding. By applying internal, external, or vacuum pressure, a cell suspension can be forced within or through a porous scaffold to establish thorough cell dispersion. Pressure can also be applied by a bioreactor system facilitating seeding and further culture of the scaffold. In case of vascularization, the vascular basal laminar network can be perfused with endothelial cells within a bioreactor system distributing the cells throughout the vessel system. It is important to put emphasis on proper single cell separation to avoid clotting of small diameter vessels [39].

2.4   Scaffold Fabrication for 3D Culture

The most widely performed 3D cultures are scaffold-based. The field of scaffold-based 3D culture relies mainly on porous scaffolds to provide a suitable in vivo like structure to mimic tissue-like structures. These scaffolds represent a template for tissue formation depicting the tissue-specific architecture and are populated with tissue-specific cell types. Additional introduced growth factors or biophysical stimuli, facilitated by bioreactor systems, might guide cellular motion, differentiation, or functionality.

The relevance of 3D culture was discussed as well as various biomaterials presented to be employed for various applications dependent on the tissue to mimic, the scientific problem, and the analytical method. The critical aspect enabling 3D cultures with the respective biomaterial is the choice of the right scaffolding technique to create the tissue-specific microenvironment and architecture. Most of the biomaterials presented above have to be processed by scaffolding techniques for the fabrication of tissue-like scaffolds to be applied for 3D cell culture improving the microenvironment of 3D cell cultures. Without proper scaffolding, most biomaterials are an unstructured mass of biopolymers. Scaffolding shapes the material's macro and microstructure. This is especially important, as the architecture of a tissue determines cellular arrangement and thereby also supports their functionality. The 3D architecture enables unique possibilities to investigate cellular behavior within their natural environment.

Irrespective of the desired tissue there are considerations for the suitability to comply with when designing a scaffold [2]. One of the most critical features of any scaffold is its biocompatibility. Cellular adhesion, viability, and function may not be compromised by, e.g., surface charge or pH. When considering implantation, potentially eliciting an immune reaction needs to be excluded to prevent encapsulation or severe inflammation. Moreover, the objective of a scaffold is to provide structural support and a template to enable the cells within to create their own natural environment, often by replacing the provided material with produced and secreted ECM. Therefore, the scaffold should exhibit a degree of biodegradability correlating to the cellular production of ECM balancing the homeostasis of production and degradation. However, the non-toxicity of the respective by-products has to be confirmed as well. Furthermore, the mechanical properties of the scaffold should resemble those of the reconstructed natural tissue that might range from 0.4 to 1500 MPa [40]. Particularly, when applying mechanical stimulation, the scaffold has to withstand the applied forces. Beyond mechanical integrity, the scaffold architecture should exhibit enough interconnected porosity facilitating nutrient/waste diffusion as well as cellular motility, ingrowth, vascularization. Furthermore, porosity enhances cellular proliferation. Nevertheless, high porosity alters mechanical properties of the scaffold. This can be balanced by using two different polymers or modifying surface energy and protein adsorption. The ideal mechanical modulus and degree of porosity is highly tissue and cell specific.

The selection of the scaffolding technique is highly dependent on the desired scaffold properties as, for example, electrospinning generates a soft tissue with high flexibility, surface area, and interconnected porosity. Whereas for a harder tissue, withstanding mechanical force and low porosity, 3D printing or freeze drying of composite materials will be suitable.

The most widely used scaffold techniques are electrospinning, freeze drying, and 3D printing [41], which can be further discriminated between 3D printing and bioprinting. Following, the already mentioned as well as further scaffolding techniques are introduced describing their field of application and their individual advantages and disadvantages.

2.4.1 Electrospinning

Electrospinning, the currently most widely used scaffolding technique due to the capacity to generate polymer matrices resembling native cellular micro and macroscale scaffold

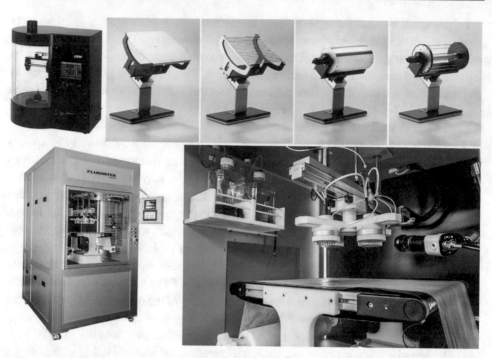

Fig. 2.5 Commercially available electrospinning systems. Upper row: Tabletop electrospinning device and different collector stages providing various platforms facilitating the fabrication of different scaffold architectures (Contipro a.s., Dolní Dobrouč, Czech Republic). Lower row: Electrospinning machine for large-scale high throughput electrospinning and electrospraying (Bioinicia, Paterna, Spain)

environment. By electrospinning, a meshwork from nano- to microscale-diameter fibers with a high surface area is generated using electrostatic attraction.

A typical electrospinning setup is comprised of three main parts: a high voltage control supply, an extrusion pump connected to a syringe with a metal tip (i.e., the spinneret), and a metallic collector. A typical setup can be self-assembled by an engineering; however, there is also a wide range of commercially available equipment (Fig. 2.5) and ready spun matrices (Fig. 2.6).

There have been more than 200 natural polymers (e.g., silk fibroin, chitosan, gelatin, collagen, etc.), synthetic polymers (e.g., PVA, PVP, PLLA, PCL, etc.), as well as composites utilized. Biological, natural, and synthetic polymers and 3D matrices thereof as well as their scaffolding by electrospinning and applications are discussed in further detail in Chap. 5.

For electrospinning, the polymers need to be homogeneously dissolved in a solvent. By charging the polymer solution in the metal nozzle with the usage of high voltage, repulsive electrostatic forces between the polymer molecules are built up. Thus, due to the attractive force between polymer solution and collector plate a steady stream of polymer solution is

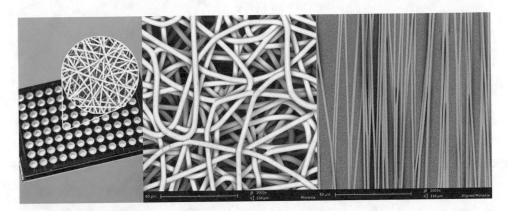

Fig. 2.6 Electrospun matrix in a 96-well plate format. Available with fibers either randomized or aligned (The Electrospinning Company Ltd., Didcot, United Kingdom)

accelerated from the nozzle towards the metal collector of opposite electrical polarity. For the generation of electrospun fibers overcoming the surface tension of the liquid polymer solution, a high voltage of 10–40 kV is required. Most of the solvent evaporates while in the air-phase during spinning onto the collector resulting in a dry polymer meshwork.

Fiber size and diameter can be adjusted by the concentration of the polymer used, the conductivity, viscosity, and surface tension of the solution, the applied voltage, the flow rate of polymer extrusion through the nozzle and the diameter of the needle, as well as the distance between needle tip and collector. Nevertheless, humidity and temperature also influence the quality and reproducibility.

Utilizing electrostatic forces, electrospinning is usually distinguished between horizontal and vertical setups, regarding the orientation of nozzle to collector, with the collector either stationary or movable.

With a rather simple setup and the adaptability of the systems, it is possible to create distinct microstructures, e.g. by the sequential use of more than one polymer pump. Moreover, despite the influence of ambient conditions, electrospinning is usually performed as a tabletop setup. Nevertheless, it should be placed within a fume cupboard as this provides ventilation to eliminate an eventually emerging strong smell and provides a safety cabinet as precaution for further lab colleagues.

An advantage of electrospinning is the high surface to volume ratio enabling the integration of active molecules for cellular guidance. Furthermore, porosity, size, and shape are easily adjustable. For example, an airflow around the spinneret can be attached. Speed and temperature of the airflow modify viscosity of the solution and evaporation of solvent, thereby effecting the morphology of the nanofibers. Electrospraying, a variation of electrospinning, accelerates droplets instead of a continuous thread onto the collector (Fig. 2.7).

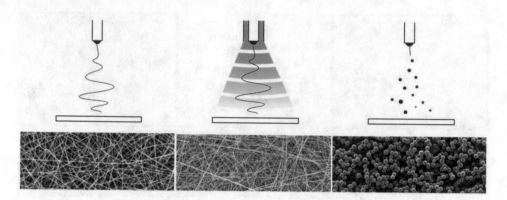

Fig. 2.7 Basic concept of an electrospinning, electroblowing, and electrospraying and their respective spun matrix (Contipro a.s., Dolní Dobrouč, Czech Republic)

Nevertheless, the mechanical properties of the electrospun scaffolds are sufficient for many TE approaches but have to be handled with care as the scaffolds are prone to mechanical disruption.

2.4.2 Bioprinting

3D printing technology resembles stacking 2D printing ink on paper. Originating from stereolithography, the technological advancement of 3D printing to stack the "print" onto the z-axis to generate 3D scaffolds was an enormous innovation. 3D printing, rapid prototyping, stereolithography, or laser sintering are used to fabricate scaffolds for 3D culture, bioreactor systems, molds, building blocks, or anything else related to biomedical applications.

The huge advantage over most other scaffolding techniques is the computer-aided design of the scaffold in advance, allowing the distinct control over scaffold parameters including location, size, and interconnectivity of pores. In contrast, for most scaffolding techniques the resulting architectural organization is composed by a certain degree of random configuration. Having distinct control over matrix generation enables the optimization of parameters such as mechanical strength, porosity, and architecture. Thereby, affecting cellular distribution, growth and nutrient supply.

An important discrimination to make aware of is between 3D printing and bioprinting as both often are used synonymously. While 3D printing comprises the fabrication of scaffolds from biomaterials for the intention to culture cells on them or the fabricating of bioreactors for cells to be cultured in, bioprinting represents printing live cells within a hydrogel ink.

For 3D printing, laser-, printing-, or nozzle-based solid free-form fabrication are methods to convert a computer designed scaffold into a polymer matrix. Similarly, to a printer, these methods generate a scaffold by moving within the x-y plane and stacking a solid 3D structure iteratively on the z-axis. To preserve the stacked architecture, printing methods use wax deposition or chemical binders to create structures out of a powder bed, whereas laser methods use electromagnetic energy to selectively polymerize or solidify a monomer solution or powdered material. Nozzle systems emit a chemically or thermally liquefied material that solidifies when extruded.

The first bioprinters were modified versions of commonly available 2D ink-based printers. In the cartridge, the ink was replaced with a polymer, and instead of printing on paper that gets pulled through the printer, the printing was performed directly on a stage with a controlled elevator to regulate the x-y-z-axis. As the setup is not very different from classical 2D printers, the accessibility for the user was high and the technology spread fast and rapidly developed further, enabling researchers to print living cells into a structured 3D tissue. Bioprinting strategies resemble those for 3D printing for material deposition and patterning using inkjet, micro-extrusion, or laser-assisted bioprinting. The difference to sole 3D printing is in laser-assisted bioprinting, cell-laden hydrogel solution is focused on a collector with focused pulses. Therefore, there is a focusing system for a pulsed laser beam, a ribbon with a donor transport support with a laser-energy-absorbing layer, and a receiving substrate facing the ribbon necessary. Inkjet bioprinter systems are customized 2D printer ejecting drops of hydrogel onto a substrate. The most common system for bioprinting is the extrusion method requiring a temperature-controlled dispensing and material-handling system, a movable stage, a fiberoptic light source for photo-initiator activation, a video camera for x-y-z command and control, and a piezoelectric humidifier. While a setup for 3D printing of materials is rather easily self-assembled, bioprinting requires the handling of live cells and therefore more specialized equipment. There are commercially available systems (Fig. 2.8) also offering bio-ink based on different materials, e.g. collagen, gelatin,

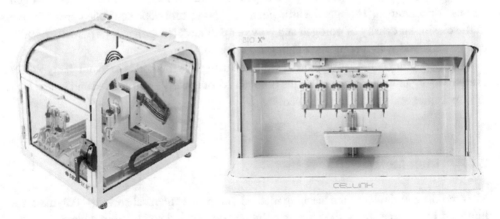

Fig. 2.8 Commercially available bioprinter systems (Brinter, Turku, Finland; Cellink AB, Gothenburg, Sweden) with multiple printing heads to apply different materials and cells in a spatial resolution

or chitosan. Multiple print heads facilitate serial dispensing of multiple materials with different cells and/or cellular concentrations.

Bioprinting, when using hydrogels, easily allows to implement cellular co-cultures, drugs, and bioactive molecules as well as their precise arrangement by printing different cell types with different cartridges. However, the implemented cells must withstand the exerted shear stress and pressure while extrusion as well as cellular viability has to be ensured for the whole time of printing. Furthermore, polymers typically used for bioprinting are alginate, collagen, or different hydrogels that exhibit high biocompatibility, but low mechanical strength. Thereby it is often necessary to also print supportive structures to aid in statics until the final complex organ structure is printed.

For more details, Chap. 11 Bioprinting is explaining the methods and applications more in depth.

More classical scaffolding techniques with similar principles and results are solvent casting/porogen leaching, melt molding, gas foaming, and freeze drying. However, they also share the same disadvantage of lacking in precise fine-tuning of the resultant scaffold microstructure, morphology, and porosity.

2.4.3   Salt Leaching

One of the oldest methods for scaffolding is solvent casting and particulate leaching. The concept of this method depends on the dispersion of solvated porogens in a polymer solution (Fig. 2.9). The liquid suspension of polymer and porogen is then solidified with the porogens dispersed throughout the solid polymer. Afterwards, this solid material is soaked in a solvent to dissolve the porogen leaving behind pores within the solid scaffold.

Pore size, density, and geometry are dependent and thereby adjustable by the porogen and its concentration. Though, the arrangement of pores cannot be controlled properly due to the dissimilar density of porogen and polymer particles. Nevertheless, a porosity of 90% can be achieved with a pore size ranging from 5 to 600μm.

A critical point in the fabrication of scaffolds using this method is to avoid the typically used toxic organic solvents or when the porogen is completely encapsulated and sealed by a tight polymer and cannot be dissolved.

2.4.4   Melt Molding

Melt molding is similar to the leaching method but avoids chemical solvents. Polymers are liquefied by heating above the melting point and cooled and solidified in a mold.

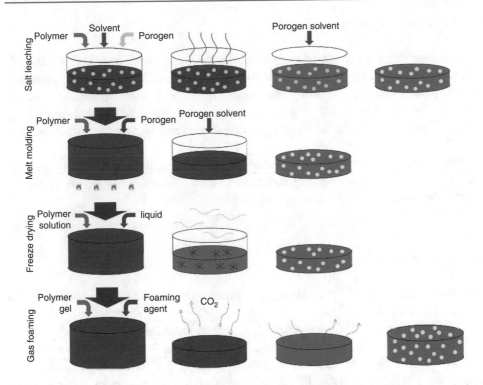

Fig. 2.9 Schematic overview of solvent casting/particulate leaching, melt molding, freeze drying, and gas foaming. Salt leaching: Dispersion of porogens in a polymer solution. Evaporation of the solvent leading to solidification while another solvent dissolves the porogens afterwards leaving behind a porous structure. Melt molding: Melting polymers with porogens under pressure leads to a solid block. The dispersed porogens can be dissolved within the block creating a porous scaffold. Freeze drying: Polymers are mixed with a liquid and frozen forming ice crystals. Dehydration of the ice crystals results in a porous scaffold. Gas foaming: A polymer solution is mixed with a foaming agent, pressed to a compact block, and gassed. When releasing the gas quickly expands foaming the material

The most used polymer for this method is poly lactic glycolic acid (PLGA) due to its low glass transition temperature. The melt molding technique can be combined with particle leaching to introduce porous structures.

By heating and melting polymers also composite materials can be generated, e.g. gelatin microspheres might be mixed with a polymer powder in a Teflon mold and then heated above the polymer glass transition temperature. The result is a composite material with polymers incorporated in the gelatin modifying its structure and characteristics (Figure SCAF).

However, despite the rather easy setup, the incorporation of bioactive molecules is restricted due to the heating.

2.4.5   Freeze Drying

A technique without the use of toxic organic solvents is freeze drying. Applying this method, an emulsion from a solvated polymer and a non-mixable liquid is homogeneously mixed, poured into a mold, and freeze dried. There, the solvents freeze and form ice crystals. The ice crystals act as porogens forming the pores within the scaffold. By dehydration under vacuum, the solvent gets evaporated resulting in a dry porous structure (Fig. 2.9).

The porosity and pore size can be manipulated by using different polymers, the ratio between polymer and emulsifying liquid, as well as controlling the growth of the ice crystals by regulation of the freezing temperature.

2.4.6   Gas Foaming

Avoiding solvents at all, gas foaming creates matrices allowing the incorporation of bioactive molecules.

For gas foaming, a foaming agent, such as sodium bicarbonate, is mixed in a polymer solution. This mixture is pressed to a solid block by high temperature compression molding and stored in a high-pressure carbon dioxide chamber for several days enabling gas infiltration inside the polymer block. Immersion of the block into an acidic aqueous solution leads to a reaction with the encapsulated foaming agent releasing gas. With the release of the entrapped gas, the scaffold foams and pores form (Fig. 2.9).

With this method, a 93% porosity with a pore size up to 100μm can be achieved. However, the foaming is hardly controllable forming a heterogeneous scaffold with an increasing porosity from a non-porous bottom of the scaffold to a highly porous top due to the rising gas. Nevertheless, after foaming, there is no residual solvent causing cytotoxicity or compromising the bioactivity of subsequently introduced molecules.

Finally, as already mentioned before, the most important scaffold property is to provide a structure to enable the cells to shape their own extracellular environment.

For scaffold-based culture systems, reproducibility between different batches is unsatisfactory.

2.4.7   Sterilization

It is critical to identify the suitable sterilization method for the biomaterial used to avoid compromising effects (Table 2.3). Some sterilization methods can alter chemical, morphological, and mechanical properties of biomaterials as sterilization processes often require harsh conditions to inactivate bacteria and pathogens [42]. In this regard, it is often advantageous to sterilize the raw materials instead of the final scaffold. This is especially

Table 2.3 Overview of different sterilization techniques [42]

Sterilization method	Advantages	Disadvantages
Autoclave	Safe; usually available	Deformation of thermoplastics
Dry heat	Safe; usually available	Deformation of thermoplastics
EtO	Minimal degradative changes; sterility easily validated	Toxic residues; surface changes; extended aeration time required; possible shrinkage of thermoplastics
Gamma	Most penetrative; crosslinking capability; sterility easily validated	Degradative alterations in polymers and bioceramics
E-beam	Shorter exposure time than gamma; sterility easily validated	Degradative changes; high cost
UV	Minimal degradative changes; inexpensive/ usually available	For surfaces only; non-FDA-approved sterilization method
Chemical	Safe; cheap	Evaporation time required

true for scaffolding techniques that directly implement cells like bioprinting as terminal sterilization processes will inevitably inactivate cells or proteins.

Heat sterilization can be performed with dry heat or steam. The latter is commonly performed in an autoclave heating water above 121 °C for at least 15 min. For proper sterilization, the steam needs to penetrate the whole material. Dry heat, on the other hand, is performed longer and at higher temperature as the heat penetration takes longer. Usually the material is stored for at least 2 h at 160 °C in a hot air oven. Whereas hot steam is detrimental for lots of materials, dry heat can also be used for powders sterilizing material before scaffolding.

Chemical sterilization by immersion in ethanol, isopropyl alcohol, formic acid, or low-concentrated peracetic acid is an option for scaffolds that are heat and radiation sensitive. However, the sterilization liquid has to be evaporated properly to not leave cytotoxic residues behind.

Instead of applying liquids, objects can also be sterilized with ethylene oxide (EtO) gas. By adding alkyl groups to sulfhydryl, hydroxyl, amino and carboxyl groups, proteins and nucleic acids get denatured. Thereby also cells and microorganisms die. However, due to the high toxicity of the EtO gas, there is a long and proper airing required to ensure safety for the operator as well as for subsequent cells seeded onto the scaffold.

Irradiation sterilization. Another method of low-temperature sterilization is gamma irradiation. The radiation penetrates through most physical barriers, being suitable for packed materials. However, the high energy electromagnetic waves do not only degrade proteins and nucleic acids, inactivate cells and microorganisms but might also cause degradative effects in some materials causing cytotoxicity.

Utilizing higher dosing rate but therefore shorter exposure time, electron beam sterilization reduces the risk for material degradation.

Ultraviolet (UV) irradiation is another electromagnetic radiation, but less penetrating than gamma irradiation and therefore used for surface sterilization.

2.5   Bioreactors Mimicking Physiological Culture Conditions

Bioreactors have emerged as favorable tool for cell culture, especially in the cell-based therapy industry. 3D structures often require additional nutrient supply due to diffusion limitations. Cells that are embedded in scaffolds have limited availability as well as exchange rate of nutrients and waste products. It has been shown that the use of bioreactors in the field of TE provides additional biomimetic stimuli, efficient delivery of nutrients and improved mixing, fluid shear stress and perfusion regimes. Consequently, formation of adequate cell–cell interactions, efficient oxygen and nutrient supply as well as continuous waste removal within the bioreactor system, enabled higher mass transfer rates, regulated cell behavior and influenced biological processes positively. Introduction of biochemical stimuli increases the functionality of scaffolds and online monitoring via incorporated sensors for pH, oxygen, and temperature, facilitate real-time feedback of culture conditions. Furthermore, enhancement of proliferation rate and reduction of a necrotic core formation in 3D scaffold has been observed under dynamic cultivation in bioreactors. Transition from static systems to bioreactors enables easy sample collection during expansion process and direct downstream analytics such as flow cytometry which aids a greater control over cultivation and facilitates efficient process optimization. These aspects display more benefits for 3D tissue engineered tissues compared to conventional 2D cell culture techniques (see also Chap. 1.3). Nevertheless, diffusional improvements within tight tissues in dynamic cultivation do not suffice. In this case, vascularization has to be integrated into the tissue model [43]. Vascularization in 3D cell culture will be addressed in Chap. 6.

Over the past decade, numerous types of bioreactors have been developed for the maintenance of a controlled microenvironment for different types of cells including red blood cells, cell lines, (mesenchymal, adult, induced pluripotent-) stem cells, and chimeric antigen receptor (CAR) T cells, to regulate appropriate cell viability, growth, differentiation and tissue development. Researchers were aiming for (1) improved standardized and reproducible processes, (2) scale-up for clinically relevant cell-based products for regenerative medicine applications, (3) superior functionality of 3D tissue grafts, and (4) establishment of in vitro models which are physiologically similar to that of in vivo tissues in order to enable pharmacological testing for various experimental parameters. Therefore, commercialization of bioreactor systems has garnered great attention as it provides a more convenient, safer, and viable method compared to traditional planar platforms (e.g., T-flasks). In spite of these benefits, regulatory requirements indicate significant challenges and specific guidelines of bioreactors for various parameters including flow rate, cell culture medium volume, requirements of different cells in such settings are limited. Further issues emerge in terms of the scale-up potential for production of industrial quantities, simplification of functional in vivo systems and control of manufacture and monitoring of miniaturized systems that reflects the complex physiology of the native tissue (see description in Chap. 1.6). Nevertheless, notable efforts have continuously been made in order to improve the clinical applicability of TE grafts by focusing on the use and optimization of biophysical stimulation and functional tissue assembly. In the following sections, two main modes of bioreactors, used for production of functional 3D tissues, will be described: mixing bioreactors (orbital shakers, spinner flasks, stirred tank, etc.) and perfusion bioreactors (hollow fiber, VITVO, fixed bed, etc.) [44, 45]. Strategies for cultivation of 3D co-cultures are further described in Chap. 8 and microfluidic systems and organ-on-the-chip approaches are explained in more detail in Chap. 10.

2.5.1   Mixing Bioreactors

One major challenge for the development of tissue-replacement grafts or stem cell-based therapy is the generation of functional cells in large quantities. Hereby, different factors including production time as well as the practical and cost-effective application, need to be considered. Mixing bioreactors provide homogenous distribution of nutrient, oxygen as well as cellular by-products during cultivation via stirring or oscillating and rocking components, while keeping the cells in suspension in the vessel. Usually, these bioreactor types are operated in combination with microcarrier technology (as described in Chap. 2. 3.1) for cultivation of adherent cells to minimize damage of cells through hydrodynamic shear forces. Hereby, optimal agitation rates need to be applied to promote proliferation and maintain viability of cells, which is challenging due to the highly dynamic and interconnected parameters of the bioreactor [46].

TubeSpin® (https://www.youtube.com/watch?v=_rEY9CUaKRk&feature=emb_title) Bioreactors (Techno Plastic Products AG (TPP), Switzerland) are disposable culture vessels which are installed on an orbital shaker to provide efficient cultivation conditions for mammalian cells (on microcarrier or as cell aggregates) in suspension. The ventilated cap allows sterile supply of oxygen via an incorporated 0.22µm sterile PTFE membrane once sealed. The conically designed bottom enables application onto standard swinging or fixed-angle rotors for cell/liquid separation by centrifugal forces. Furthermore, it facilitates easy harvest of cells within the tubes due to its low-adherent conical bottom avoiding additional transfer step. The available formats allow working volumes from 10, 35 up to 400 ml. A successful high yield expansion as well as chondrogenic differentiation in a pellet culture of human stem cells in the TubeSpin® bioreactor has been reported in literature [47, 48]. These studies demonstrated the efficiency of the TubeSpin® bioreactor 600 for cell culture applications as it offers great possibilities for the optimization of suspension cell culture parameters and demonstrates simple and cost-efficient scale-up opportunities.

CERO (OLS Omni Life Sciences GmbH & Co. KG, Germany) is a bench-top incubator-bioreactor hybrid, which uses similar mixing principles for suspension cell culture as the aforementioned bioreactor. Contrary to the oscillating movement of the shaker used for TubeSpin®, the tubes are placed in the CERO inserts and facilitate axial rotation induced by an incorporated rotor, which can be set at different speed for each individual tube.

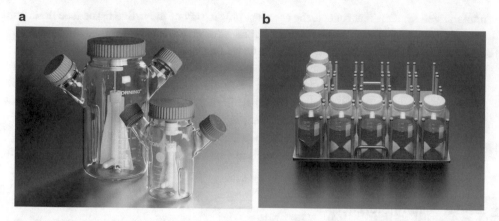

Fig. 2.10 Exemplary mixing bioreactors for cell culture. (**a**) 15 L Corning® reusable glass spinner flask with 100 mm flat center cap and 2 angled sidearms and (**b**) TubeSpin® Bioreactor 600 with a working volume of 400 ml

Furthermore, the CERO provides ultra-low attachment surfaces of its tubes and has been reported to generate reproducible spheroids using cell lines, primary cells, tissue pieces, and organoids. Additionally, CERO allows long-term cultivation and supports cell viability and differentiation. Nonetheless, it has also been reported that spheroids generated from hepatic cell lines resulted in unstable spheroids and the formation of necrotic core during a week of cultivation. Also, further challenges indicated to be the labor-intensive aspect, susceptibility to contaminations, and spatial and time limitations. Further optimization of cultivation parameters such as rotation speed needs to be made for each cell type to obtain anticipated results in the bioreactor (https://www.youtube.com/watch?v=9pSPhGUgT1s; Link zu youtube Video/Bonusmaterial für online).

Corning® Disposable Spinner Flasks (Corning Inc., USA) enable enhanced mass transfer by incorporation of a stirring element such as magnetic stirrer at the bottom, which creates a convective flow and produces hydrodynamic forces. It consists of a cylindrical glass vessel, where cell aggregates or cells seeded on microcarriers are kept in suspension during cultivation (Fig. 2.10). Spinner flasks can be conducted in batch, fed-batch, or either continuous culture mode. Cellular distribution, viability, and differentiation capacity has been improved through the mixing regimen of the spinner flask bioreactor. It has found applications in bone TE but has been reported to only promote ECM production at the surface of scaffolds and increased turbulence-collision on the scaffold through mixing, affecting cell growth and tissue formation.

Stirred tank bioreactors (*STBR*) such as Ambr® 250 modular (Sartorius AG, Germany), Mobius® CellReady 3 L (Merck Millipore, USA), and BioBLU (Eppendorf Austria GmBH, Austria) have been known as the classical bioreactor systems and represent the predominant systems for large- scale clinical grad expansion of MSC [49] and especially established protocols are available for the biopharmaceutical production of antibiotics from predominantly recombinant Escherichia coli (E. coli) or Chinese hamster ovary (CHO) cell lines. STBR are offered as stainless-steel systems with working volumes ranging from 2 to 1000 L as well as re-useable systems as flexible bags or rigid vessels with volumes of up to 2000 L. Whereas, smaller STBR (5–200 ml) are mainly used for research or process development purposes. Besides the flexibility in working volumes, STBR are also available in a variety of designs and distinct properties, which could result in different biological performances for each system. Cultivation of human MSC at maximum scale has been achieved in 50 L STBR involving microcarrier technology or cell aggregate culture [50].

The success of the process of individual STBR highly depends on the varying experimental conditions and procedures. Central process engineering properties of bioreactors include vessel geometry, vessel material, aeration, and impeller geometry. Depending on the specific application and the volume size these process properties, especially the impeller design, have a great impact on mixing time, power input, volumetric mass transfer coefficient, and formation of shear gradients. Therefore, the choice of an appropriate agitator such as axial and radial flow impellers is crucial for the outcome of the cultivation of specific types of microorganism, human or animal cells. Therefore, the agitation rate needs to be optimized for the specific culture as homogenous mixing is necessary to prevent the formation of substrate gradients or cell-loaded microcarrier or cell aggregate cluster due to cell bridging. Equally important indicates the aeration in the system, which needs to be controlled in order to optimize the mixing process. High aeration in combination with high agitation ensures proper mixing but causes increased shear stress. On the other hand, too high agitation rate can cause disruption of cell–cell contact or cell–adhesion to the microcarrier or afflict too high shear stress on cell and can damage cells. For a reliable comparison of the physical capability of different STBR systems, standardized protocols for characterization of such bioreactors need to be established, which could further support process optimization and guide scale transfer for industrial production. Moreover, the assessment of the biological capability of the bioreactor must also be evaluated in order to determine if the system is suitable for the specific cultivation. Ultimately, the choice out of all the available bioreactor systems needs to be made according to the requirements of the desired process and application [51].

Vertical-Wheel™ (https://www.pbsbiotech.com/vertical-wheel.html) bioreactor (PBS Biotech Inc., USA) is a single-use vessel which consists of a vertically oriented impeller and further key features include a long wheel radius, peripheral paddles and rounded, ultra-low adherence vessel bottom, and oppositely oriented internal vanes (Fig. 2.11). All these properties contribute for a homogenous and gentle mixing as well as a tangential flow through the combination of radial and axial flow impeller. The wheel can be operated via buoyant force of gas bubbles introduced into the bioreactor (AirDrive) or magnetic coupling (MagDrive) and is enclosed freely around a stationary in the vessel. These agitation mechanisms provide mixing and particle suspension with low power input as well as agitation speed, which in turn demonstrate to be more beneficial for sensitive cell types such as MSC in terms of improved cell attachment and growth. The Vertical-Wheel™ bioreactor is available as lab-scale vessels (0.1–0.5 L) up to larger production

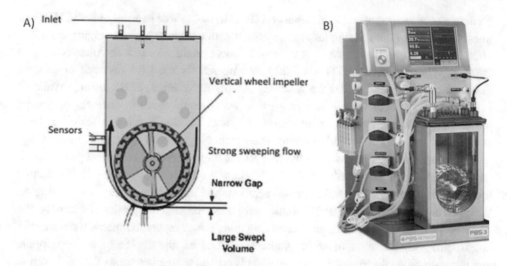

Fig. 2.11 Single-use vertical-wheel™ bioreactor. (**a**) Key features of the vertical-wheel bioreactor including: U-shape vessel, oppositely oriented axial vanes, sizeable impeller zone and vertical-wheel impeller. (**b**) Representative large-scale cell manufacturing model: MagDrive PBS 3-L bench-top bioreactor

units (up to 500 L) and showed maintenance of homogenous mixing properties. In addition, reduction of long operation times between batches can be avoided due to its single-use application as well as prevent cross-contamination and has been approved to be cGMP-compliant. Initial studies have reported microcarrier-based cultivation of human MSC and recently also of human induced-pluripotent stem cells (hiPSC). Rodrigues et al. [52] were able to obtain an approximate seven-fold increase expansion of hiPSC in 6 days using the Vertical-Wheel™ PBS 0.5, recombinant vitronectin-coated microcarrier and xeno-free reagents, suggesting the suitability of this system for cell-based bioprocesses.

Xuri™ (https://www.youtube.com/watch?v=jn972ZqCQ-I) *W25 cell expansion system* (Cytiva, USA) relies on a rocking platform which induces a wave motion of the suspension inside a gas permeable culture bag (1–100 L) in order to achieve a homogenous mixing and bubble-free aeration. Usually, gentle rocking regimes are applied to provide sufficient oxygen transfer. However, similar to all bioreactors all process parameters need to be adapted to the specific application including cell type and desired product. They have been widely used for preclinical research purposes, seed trains for larger production-scale cultures, and vaccine production due to their disposable technology.

2.5.2 Perfusion Bioreactors

Compared to mixing bioreactors described in Chap. 3.5.1, in perfusion bioreactors a flow of cell culture medium is applied upon the cell population to distribute the oxygen and nutrients throughout the entire bioreactor vessel and porous scaffolds. Furthermore, it has

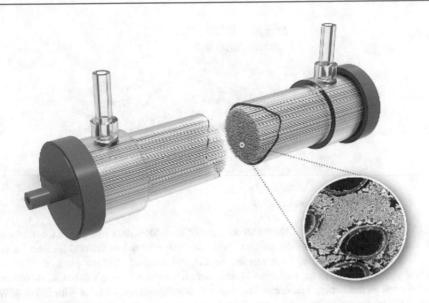

Fig. 2.12 *Cross-section of a medium cartridge of a hollow fiber bioreactor (FiberCell Systems Inc., USA).* This module is available in both low (5 kDa) and high (20 kDa) MWCO. Furthermore, it offers a surface area of up to 4000 cm^2, which can be compared to the area of 22 T-175 cell culture-treated flasks. The cross-section indicates seeded cells on the outer surface of individual hollow fiber, which have grown in high density during cultivation

also been used to apply controlled shear stress to induce MSC differentiation or enhance release of by-products such as extracellular vesicles. Most important key parameter is the flow rate, which is therefore required to be optimized specifically for each bioreactor setup and cell type to prevent insufficient oxygen and nutrient supply to the cells. Various types and versatile configuration settings are available commercially or are developed in-house for numerous biomedical applications. In the next sections, a few examples of perfusion bioreactors will be described.

Hollow fiber bioreactor (FiberCell systems Inc., USA) (KD Bio (https://www.youtube.com/watch?v=uSvfrDAKXok, https://www.kdbio.com/video-animation-how-does-a-hollow-fiber-bioreactor-work/) S.A.S, France) is composed of a bundle of parallel, semi-permeable hollow fibers (HF) out polysulfone assembled in a cylindrical cartridge (Fig. 2.12). Different sizes of HF modules are available providing surface areas ranging from 80 m^2 up to 1.2 m^2.

Membrane properties including pore structure and permeability can be highly controlled via the manufacturing process. HF bioreactors are offered with a variety of properties where the pore morphology and size can be tailored as a selective barrier for target molecules which are hindered of diffusion due to the given pore dimension, while continuous nutrient and waste product transport is aided in the liquid phase. HF are classified according to the defined molecular weight cut-off (MWCO) ranging from 6 up to 190 kDa by the manufacturer. Hereby, 90% of molecules in the system, indicating a

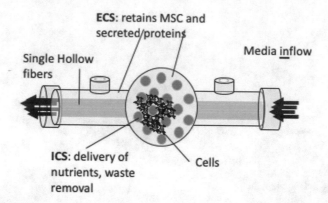

Fig. 2.13 Illustrative scheme of the inner architecture of the hollow fiber bioreactor. Continuous cell growth can be facilitated by seeding adherent cells onto the external or internal lumen of a single hollow fiber. Moreover, constant exchange of nutrient and removal of waste products is provided during the entire cultivation by perfusion of cell culture medium through the ICS. Additionally, molecules such as exosomes, monoclonal antibodies, or recombinant proteins with larger MWCO than the hollow fiber can be retained and be highly concentrated in the ECS. Abbreviations: ECS = extracapillary space, ICS = intracellular space

larger molecular weight than the membrane, are retained. Subsequently, selective passage is enabled through the membranes' MWCO and therefore HF bioreactors have been commonly used in order to prevent exchange of immune-competent species within the system or isolate secreted by-products such as extracellular vesicles. Besides the tailored MWCO of the membrane, HF bioreactors allow multiple medium flow regimes and cell seeding options: medium flow through (1) extracapillary space (ECS) and (2) intracapillary space as well as seeding of cells (3) on the external surface and (4) internal lumen of HFs (as illustrated in Fig. 2.13). Choosing a configuration of HF bioreactor setting, where medium flow and cells are directly exposed to each other, will permit cells to be subjected to shear stress and nutrient supply is facilitated by convective transport. Another option is to create a co-culture model within the system, by seeding different cell types on the outer surface as well as in the internal lumen of HFs as it was used for endothelial cells under flow studies in order to evaluate blood stream characteristics. Additional applications are mass expansion of MSC for cell-based therapies (Quantum expansion cell system, Terumo BCT, USA), monoclonal antibody, recombinant proteins, adenoviral vectors and exosomes production, as well as in vitro toxicology studies of chemicals and drugs, and 3D cell culture platform to mimic tissue-like densities in long-term cultivations such as the human gut [53, 54]. This however indicates the requirement for optimization of process parameters (e.g., perfusion rate, fluid properties), regulation of mass transport parameters (e.g., oxygen, nutrient, and waste molecules), and culture conditions (e.g., cell type, feeding strategy, and seeding density) which need to be adapted and evaluated for each application as success of the system is highly dependent on these different operating parameters.

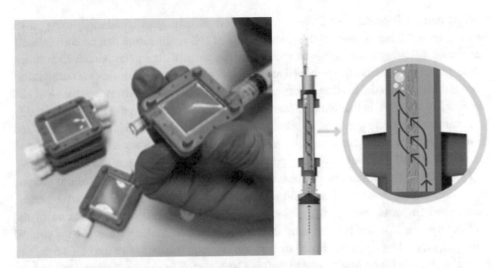

Fig. 2.14 VITVO® technology. Portable and small VITVO® device ensures safe and easy transport from one lab to another. Furthermore, two chambers are separated by a 3D matrix and simultaneously serve as a filter to facilitate retainment of cells and successful colonization

For many years, it has been a challenge to overcome the diffusive limitation in different systems to deliver nutrients sufficiently and consequently form hypoxic and necrotic conditions for cells. This system has promising potential to increase the ability to design a microenvironment for cells in vitro that mimic the physiological conditions of living tissue. Especially, due to various benefits such as 3D platform, compactness, excellent mass transfer properties, large surface area and cell protection from mechanical stresses that are yielded by its HF geometry [55].

VITVO® (https://rigenerand.it/vitvo/) (Rigenerand, Italy) is a novel bioreactor which creates a miniaturized 3D in vitro and in vivo-like microenvironment for drug screening and cell expansion purposes in a closed system. It ensures easy handling, time-saving factors, a minimized risk of contamination, portable properties due to its small size (as shown in Fig. 2.14) and convenient monitoring over time. VITVO® consists of a perimetral frame with two transparent oxygenation membranes, allowing gas exchange and visibility during cultivation. The inner space is separated by a fiber-based matrix (thickness: 400μm) composed of an inert and biocompatible synthetic polyester and its dry volume represents approximately 90% of the entire VITVO volume. Consequently, the 3D matrix creates two completely separate chambers for cell seeding and medium flow accessible through an inlet and outlet (Fig. 2.14). The device is ready-to-use and liquid is initially passed through one chamber first and once filled the second chamber is entered. This permits cell seeding of the 3D matrix in a controlled manner as the 3D matrix acts as a filter and retains cells and colonization on fibers. Cell suspension is easily injected using a syringe connected to the inlet port and medium change is also rapidly conducted in the same way. Furthermore, if cells are pre-stained with a fluorescence dye, it can be directly viewed under a fluorescence microscope to observe cellular growth and morphology after loading onto the matrix. Real-time quantification of cell proliferation can be conveniently performed by using either luminescence or fluorescence with a plate reader. Due to the small, portable size and self-contained technology of the bioreactor, a safe and straightforward shipment from the loading laboratory to another laboratory, performing read-out and histological analyses, is facilitated. Moreover, the simplicity of design ensures adaptability to common laboratory equipment which benefits academic and industrial research and development fields. Efficient gas and nutrient exchange by diffusion and high yield of viable culture cells as well as the compatibility with various cell types has been reported. VITVO offers the optimal compromise between the reduction of culture medium volume and a successful creation of a tissue structure closely mimicking the in vivo setting, facilitates comparable results with in vivo studies. The predominant application of the VITVO system is in the field of drug development and toxicology, relevant for clinical settings as it enables rapid functional drug screening tests and evaluates the cell's behavior for a possible prediction of a patient response. Candini et al. have investigated primary lung cancer cells in the VITVO® and observed the TIL activation caused by nivolumab against the tumor cells. As a consequence, an anti-tumor response was provoked due to anti-PD1 antibodies and increased tumor antigen expression was observed, caused by 3D cell growth. Herein, several chemotherapy, biologics, and cell-based anti-cancer agents were evaluated for their efficacy and compared to in vivo preclinical xenogenic transplant models. VITVO® is an innovative 3D in vitro model and demonstrates promising applications especially for preclinical investigations in oncology and could expand to further relevant areas such as toxicology [56].

2.5.3 Special Types of Bioreactors

CelCradle™ (https://www.youtube.com/watch?v=SmLh4LC02pg) (Esco Lifesciences GmbH, Germany) is a bench-top bioreactor and consists of a vessel filled with BioNOC™ II carriers (5.5 g; 13,200 cm^2 surface area for cell attachment and growth) in order to provide a packed bed inside the reactor (as depicted in Fig. 2.15). In addition, CelCradle™ can be converted from a lab-scale culture to large-scale setting by direct multiplication of bottles and demonstrates reduced labor-intensity. Cells are seeded onto these structures while medium is perfused in a tide motion principle as medium is perfused alternately by decompression and compression of the bottom part of the CelCradle™. Hereby, elimination of foaming, constant vertical oscillation, and low shear at all scales is created, avoiding shear peaks (eddy size > ~65% of the microcarrier size), as observed in an STR near the impellers, to enhance cell growth and promote migration into the bed. Furthermore, alternating exposure to aeration and nutrition provides a dynamic interface between the culture medium, air and cell surface which enhances nutrition levels and produces high density culture of adherent cells. Successful expansion of cell lines including HEK-293, BHK-21, and MDCK in the CelCradle™ bioreactor has been reported. Packed bed

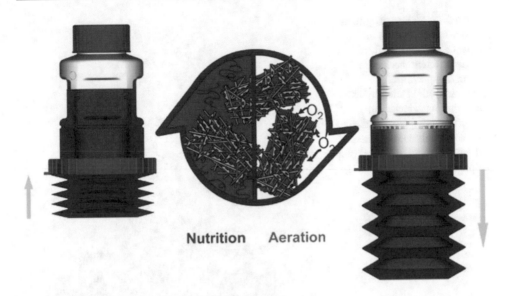

Nutrition   Aeration

Fig. 2.15 Tide motion principle of the CelCradle™ bioreactor. Bottom part of the reactor which contains the cell culture medium is alternately oscillated in down and up motion. As a consequence, exposure of the cells to medium and air refers to the time of the oscillation in the respective direction. This creates enhanced dynamic conditions within the reactor and increased aeration and nutritional exchange from the medium and air to the cells. Image obtained from Esco Aster webpage (http://www.escoaster.com/tide-technology/CelCradleTM-Lab-Scale-Adherent-Bioreactor/)

bioreactors are predominantly used for the culture of anchorage-dependent or adherent cells, production of human and animal vaccines and research purposes.

In terms of differentiation capabilities, MSC seeded on scaffolds in packed bed bioreactors have resulted in successful osteogenic differentiation due to the physiological environment and additional differentiation stimulus provided by the surface stiffness and structure of the scaffold as well as the flow properties of the reactor. However, notable efforts have been made to confer the challenges regarding the generation of a homogenous environment, easy detachment of cells during harvest, and scalability of the bioreactor. Still most studies have reported loss of viability of around 20% caused by the stress during harvest, which still needs to be addressed further [57].

TC-3 (https://www.youtube.com/watch?v=9FF3_hjHj2Y, EBERS Medical Technology. 2013; TC-3 bioreactor mechanical stimulation for stem cell culture. Retrieved September 28, 2013, from http://www.ebersmedical.com/files/brochures/Brochure_TC-3.pdf) bioreactor (Ebers Medical Technology, Spain) displays simple, easy-to-use, and multi-purpose cell culture bioreactor that provides user-defined axial loading regimes including tension and compression for mechanical stimulation of cells on tissues or scaffolds. Porous and cylindrical scaffolds, membranes, and sheet-like scaffolds can be both horizontally and vertically fixed. Additionally, samples can be immersed in an air–liquid interface and enable long-term cultivation under mechanical stimuli (as seen in Fig. 2.16). Generally,

Fig. 2.16 TC-3 loaded systems. This bioreactor consists of 3 separated chambers which can provide different mechanical loading on specimen. Mechanical stimulation by axial loading regimes including tension and compression is possible or only measurements of the mechanical properties of a sample or tissue

the TC-3 is constructed in a simplified way, allowing convenient connection and disconnection between the reactor and the rest of the control unit system. Furthermore, the chamber can be easily removed from the placement platform with the option to maintain a desired deformation state of the sample for microscopic observations. Finally, the lightweight reactor fits and is compatible with most of the standard incubators and sterilization of parts is conducted via autoclave, which enables adaptability to common laboratories [58].

For various differentiation procedure especially chondrogenic and osteogenic differentiation, mechanical stimuli such as cyclic compression or strain has demonstrated a beneficial factor by modulating the biosynthetic activity of the cells to direct specific lineage differentiation. Moreover, MSC reveal to be extremely sensitive to fluid shear stress and strain, which could be beneficial for expansion and differentiation processes. Therefore, the desire of researchers for bioreactors offering mechanical stimulation garnered great attention for the recreation of functional TE in vitro constructs that withstand similar mechanical loads as in vivo. TC-3 displays an ideal platform compared to traditional testing machines and permit scale-up processes.

In summary, the trend in cell culture is moving more and more towards a dynamic setting as it provides numerous benefits described in Sect. 2.2 and Chap. 3.5 in order to enhance cell viability, growth, and differentiation compared to static cultivation. The different designs and modes of bioreactor systems offer a broad versatility, which could further be optimized and modulated for large-scale production and biomedical applications. The demand for safe, reliable, and effective alternatives to animal models is rising and the attention for in vitro models increased in research. As an example, a fully automated skin factory for the production of human tissue equivalents has been implemented by an interdisciplinary group of Fraunhofer scientists (https://www.selectscience.net/ SelectScience-TV/Videos/the-skin-factory–automated-tissue-culture/?videoID= 2434). This skin factory (Fig. 2.17) enables a continuous process chain for tissue engineered skin starting with the extraction of cells from a human skin biopsy up until a functional tissue graft. The production plant is divided into four modes: (1) Cell extraction module, (2) Cell expansion module, (3) Tissue cultivation module, and (4) Quality control. Herein, these human tissue equivalents can be employed for testing the safety of new biochemical agents which are supposed to be applied as a therapy to the human body. Even though the concept sounds straightforward, it demonstrates expensive, labor-intensive, and complicated processes for manufacturing. Along with other TE bioreactor systems such as previously mentioned in this chapter still display several practical barriers. Therefore, further advancement of technologies and custom-designed systems for TE is anticipated in order to meet the requirements for clinical translation.

Take-Home Messages
- The trend of using 3D in vitro models constantly progresses as creating a more physiological environment for cells in vitro demonstrates great impact towards cellular behavior and ultimately increases the relevance for translation of the gained results into clinical applications.
- Biomimetic materials (3D scaffolds) offer a structure for cells, especially for anchorage-dependent cells, to attach and grow on a substrate mimicking their natural environment.
- The scaffold's properties can be defined during scaffolding fabrication, offering versatile implementations of a multitude of materials.
- Bioreactors are considered favorable tools for cell culture as it facilitates efficient nutrients supply and exchange, continuous waste removal as well as enabling the implementation of various stimuli.
- Key components of 3D culture including cell type, biomaterial, and bioreactor systems need to be carefully selected and evaluated towards each other.
- Development and optimization of 3D cell culture equipment need to be adapted to the requirements of the specific application as well as to allow for an easy handling for the applicant.

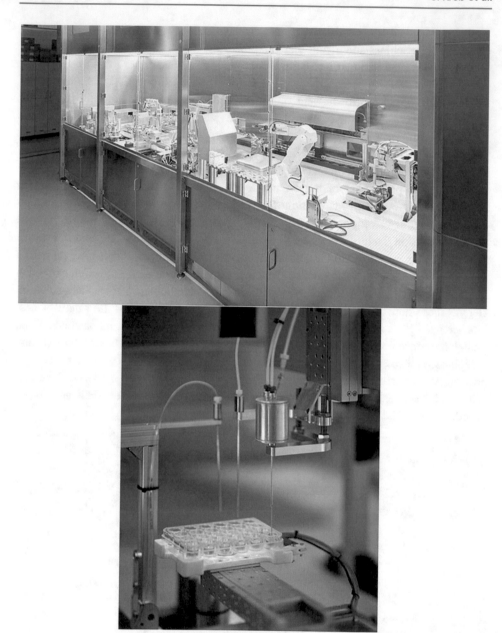

Fig. 2.17 Tissue factory. Automated production of human skin equivalents in one single production plant for large and standardized production of tissue engineered skin constructs

References

1. Ravaglioli A, Krajewski A. Antonio Ravaglioli. Netherlands: Springer; 1992.
2. O'Brien FJ. Biomaterials & scaffolds for tissue engineering. Mater Today. 2011;14(3):88–95.
3. Ratner B, Hoffman A, Scheon F, Lemons J. Biomaterials science: an introduction to materials in medicine. 3rd ed. Waltham, MA: Academic Press; 2013.
4. Engler A, Sen S, Sweeney HL, Discher DE. Matrix elasticity directs stem cell lineage specification. Cell. 2006;126(4):677–89.
5. Dolcimascolo A, Calabrese G, Conoci S, Parenti R. Innovative biomaterials for tissue engineering. London, UK: IntechOpen; 2019.
6. Hutmacher D. Scaffolds in tissue engineering bone and cartilage. Biomaterials. 2000;21 (24):175–89.
7. Haslik W, Kamolz L-P, Nathschläger G, Andel H, Meissl G, Frey M. First experiences with the collagen-elastin matrix Matriderm® as a dermal substitute in severe burn injuries of the hand. Burns. 2007;33(3):364–8.
8. Baino F, Hamzehlou S, Kargozar S. Bioactive glasses: where are we and where are we going? J Funct Biomater. 2018;9(1):25.
9. Egger D, Oliveira C, Mallinger B, Hemeda H, Charwat V, Kasper C, Egger D, Oliveira AC, Mallinger B, et al. From 3D to 3D: isolation of mesenchymal stem/stromal cells into a three-dimensional human platelet lysate matrix. Stem Cell Res Ther. 2019;10:248.
10. Choi M, Choi D, Hong J. Multilayered controlled drug release silk fibroin nano-film by manipulating secondary structure. Biomacromolecules. 2018;19(7):3096–103.
11. Pandey S, Rathore K, Johnson J, Cekanova M. Aligned nanofiber material supports cell growth and increases osteogenesis in canine adipose-derived mesenchymal stem cells in vitro. J Biomed Mater Res A. 2018;106(7):1780–8.
12. Sawa Y, Miyagawa S, Sakaguchi T, Fujita T, Matsuyama A, Saito A, Shimizu T, Okano T. Tissue-engineered myoblast sheets improved cardiac function sufficiently to dis-continue LVAS in a patient with DCM: report of a case. Surg Today. 2012;42(2):181.
13. Sato MM. Articular cartilage regeneration using cell sheet technology. Anat Rec. 2014;297:36–43.
14. Fujita H, Shimizu K, Nagamori E. Application of a cell sheet-polymer film complex with temperature sensitivity for increased mechanical strength and cell alignment capability. Biotechnol Bioeng. 2009;103:370–7.
15. Hirose M, Kwon O, Yamato M, Kikuchi A, Okano T. Creation of designed shape cell sheets that are noninvasively harvested and moved onto another surface. Biomacromolecules. 2000;1 (3):377–81.
16. Haraguchi Y, Kagawa Y, Hasegawa A, Kubo H, Shimizu T. Rapid fabrication of detachable three-dimensional tissues by layering of cell sheets with heating centrifuge. Biotechnol Prog. 2018;34(3):692–701.
17. Tan K, Teo KL, Lim JF, Chen AK, Reuveny S, Oh SK. Serum- free media formulations are cell line-specific and require optimization for microcarrier culture. Cytotherapy. 2015;17:1152–65.
18. Heathman TR, Glyn VA, Picken A, Rafiq QA, Coopman K, Nienow AW, Kara B, Hewitt CJ. Expansion, harvest and cryopreservation of human mesenchymal stem cells in a serum-free microcarrier process. Biotechnol Bioeng. 2015;112(8):1696–707.
19. Goh TK-P, Zhang Z-Y, Chen AK-L, Reuveny S, Choolani M, Chan JKY, Oh SK-W. Microcarrier culture for efficient expansion and osteogenic differentiation of human fetal mesenchymal stem cells. Biores Open Access. 2013;2:84–97.

20. Frauenschuh S, Reichmann E, Ibold Y, Goetz P, Sittinger M, Ringe J. A microcarrier-based cultivation system for expansion of primary mesenchymal stem cells. Biotechnol Prog. 2007;23 (1):187–93.

21. Li B, Wang X, Wang Y, Gou W, Yuan X, Peng J, Guo Q, Lu S. Past, present, and future of microcarrier-based tissue engineering. J Orthop Translat. 2015;3(2):51–7.

22. Pereira Chilima TD, Moncaubeig F, Farid SS. Impact of allogeneic stem cell manufacturing decisions on cost of goods, process robustness and reimbursement. Biochem Eng J. 2018;137:132–51.

23. Kuan-Liang A, Chen X, Choo AB, Reuveny S, Oh SK. Critical microcarrier properties affecting the expansion of undifferentiated human embryonic stem cells. Stem Cell Res. 2011;7(2):97–111.

24. Pettersson S, Wetterö J, Tengvall P, Kratz G. Cell expansion of human articular chondrocytes on macroporous gelatine scaffolds-impact of microcarrier selection on cell proliferation. Biomed Mater. 2011;6:065001.

25. Malda J, Van Blitterswijk CA, Grojec M, Martens DE, Tramper J, Riesle J. Expansion of bovine chondrocytes on microcarriers enhances redifferentiation. Tissue Eng. 2003;9:939–48.

26. Drury JL, Mooney DJ. Hydrogels for tissue engineering: scaffold design variables and applications. Biomaterials. 2003;24(24):4337–51.

27. Dhawan S, Varma M, Sinha VR, Dhawan K. High molecular weight poly(ethylene oxide)-based drug delivery systems - part I: hydrogels and hydrophilic matrix systems. Pharm Technol. 2005;72:72–80.

28. Ma L, Deng L, Chen J. Applications of poly(ethylene oxide) in controlled release tablet systems: a review. Drug Dev Ind Pharm. 2013;40(7):845–51.

29. Nikolova MP, Chavali MS. Recent advances in biomaterials for 3D scaffolds: a review. Bioactive Mater. 2019;4:271–92.

30. Yang G, Xiao Z, Long H, Ma K, Zhang J, Ren X, Zhang J. Assessment of the characteristics and biocompatibility of gelatin sponge scaffolds prepared by various crosslinking methods. Sci Rep. 2018;8(1):1616.

31. Smith LA, Liu X, Ma PX, Smith LA, Liu X, Ma PX. Tissue engineering with nano-fibrous scaffolds. Soft Matter. 2008;4(11):2144–214-9.

32. Gao C, Deng Y, Feng P, Mao Z, Li P, Yang B, Deng J, Cao Y, Shuai C, Peng S. Current Progress in bioactive ceramic scaffolds for bone repair and regeneration. Int J Mol Sci. 2014;15:4714–32.

33. Hench LL, Splinter RJ, Allen W, Greenlee T. Bonding mechanisms at the interface of ceramic prosthetic materials. J Biomed Mater Res. 1971;5:117–41.

34. Sekiya N, Ichioka S, Terada D, Tsuchiya S, Koboyashi H. Efficacy of a poly glycolic acid [PGA]/ collagen composite nanofibre scaffold on cell migration and neovascularization in vivo skin defect model. J Plast Surg Hand Surg. 2013;47(6):498–502.

35. Yu X, Tang X, Gohil S, Laurencin C. Biomaterials for bone regenerative engineering. Adv Healthc Mater. 2015;4(9):1268–85.

36. Porzionato A, Stocco E, Barbon S, Grandi F, Macchi V, De Caro R. Tissue-engineered grafts from human Decellularized extracellular matrices: a systematic review and future perspectives. Int J Mol Sci. 2018;19(14):4177.

37. Gilpin A, Yang Y. Decellularization strategies for regenerative medicine: from processing techniques to applications. Biomed Red Int. 2017;2017:1–13.

38. Lee JS, Choi YS, Cho S-W. Decellularized tissue matrix for stem cell and tissue engineering. Adv Exp Med Biol. 2018;1064:161–80.

39. Villalona GA, Udelsman B, Duncan DR, McGillicuddy E, Sawh-Martinez RF, Hibino N, Painter C, Mirensky T, Erickson B, Shinoka T, Breuer CK. Cell-seeding techniques in vascular tissue engineering. Tissue Eng Part B Rev. 2010;16(3):341–50.

40. Hollister SJ. Porous scaffold design for tissue engineering. Nat Mater. 2005;4(7):518–24.

41. Mabrouk M, Beherei HH, Das DB. Recent progress in the fabrication techniques of 3D scaffolds for tissue engineering. Mater Sci Eng C. 2020;110:110716.
42. Baume AS, Boughton PC, Coleman NV, Ruys AJ. Sterilization of tissue scaffolds. In: Characterisation and design of tissue scaffolds. Sawston, UK: Woodhead Publishing; 2016. p. 225–44.
43. Godara P, Mc Farland CD, Nordon RE. Nordon design of bioreactors for mesenchymal stem cell tissue engineering. J Chem Technol Biotechnol. 2008;83:408–20.
44. Stephenson M, Grayson W, Stephenson M, Grayson W. Recent advances in bioreactors for cell-based therapies. F1000 Res. 2018;7:F1000.
45. Schnitzler AC, Verma A, Kehoe DE, Jing D, Murrell JR, Der KA, Aysola M, Rapieko PJ, Punreddy S, Rook MS. Bioprocessing of human mesenchymal stem/stromal cells for therapeutic use: current technologies and challenges. Biochem Eng J. 2016;108:3–13.
46. Ahmed S, Chauhan VM, Ghaemmaghami AG, Aylott JW. New generation of bioreactors that advance extracellular matrix modelling and tissue engineering. Biotechnol Lett. 2019;41:1–25.
47. Rbia N, Bulstra L, Bishop A, van Wijnen A, Shin A. A simple dynamic strategy to deliver stem cells to Decellularized nerve allografts. Plast Reconstr Surg. 2018;142(2):402–13.
48. Pirrone C, Gobbetti A, Caprara C, Bernardini G, Gornati R, Soldati G. Chondrogenic potential of hASCs expanded in flask or in a hollow-fiber bioreactor. J Stem Cell Res Med. 2017;2(2):1–10.
49. Sart S, Agathos SN. Large-scale expansion and differentiation of Mesenchymal stem cells in microcarrier-based stirred bioreactors. Bioreact Stem Cell Biol. 2016;1502:87–102.
50. Lawson T, Kehoe D, Schnitzler A, Rapiejko P, Der K, et al. Process development for expansion of human mesenchymal stromal cells in a 50 L single-use stirred tank bioreactor. Biochem Eng J. 2017;120:49–62.
51. Rafiq QA, Coopman K, Hewitt CJ. Scale-up of human mesenchymal stem cell culture: current technologies and future challenges. Curr Opin Chem Eng. 2013;1(2):8–16.
52. Rodrigues CA, Silva t P, Nogueira DE, Fernandes TG, Hashimura Y, Wesselschmidt R, Diogo MM, Lee B, Cabral JM. Scalable culture of human induced pluripotent cells on microcarriers under xeno-free conditions using single-use vertical-wheel™ bioreactors. J Chem Technol Biotechnol. 2018;12(93):3597–606.
53. Lechanteur C, Baila S, Janssens ME, Giet O, Briquet A, Baudoux E, Beguin Y. Large-scale clinical expansion of Mesenchymal stem cells in the GMP-compliant, closed automated quantum® cell expansion system: comparison with expansion in traditional T-flasks. J Stem Cell Res Therapy. 2014;8:4.
54. Yan L, Wu X. Exosomes produced from 3D cultures of umbilical cord mesenchymal stem cells in a hollow-fiber bioreactor show improved osteochondral regeneration activity. Cell Biol Toxicol. 2020;36:165–78.
55. Eghbali H, Nava MM, Mohebbi-Kalhori D, Raimondi MT. Hollow fiber bioreactor technology for tissue engineering applications. Int J Artif Organs. 2016;39(1):1–15.
56. Candini O, Grisendi G, Foppiani EM, Brogli M, Aramini B, Masciale V, Spano C, Petrachi T, Veronesi E, Conte P, Mari G, Dominici M. A novel 3D in vitro platform for pre-clinical investigations in drug testing, gene therapy, and Immuno-oncology. Sci Rep. 2019;9(1):7154.
57. Wu YY, Yong D, Naing MW. Automated cell expansion: trends and outlook of critical technologies. Cell Gene Therapy Insights. 2018;4(9):843–863.
58. Landau S, Ben-Shaul S, Levenberg S. Oscillatory strain promotes vessel stabilization and alignment through fibroblast YAP-mediated Mechanosensitivity. Adv Sci. 2018;5(9):180056.

A View from the Cellular Perspective

Janina Burk

3

Contents

What You Will Learn in This Chapter

This chapter aims to convey a tangible impression of how cell culture procedures affect the cultured cells. Many standard cell culture procedures have been developed decades ago and are still being used routinely. However, they may to some extent be harmful to the cultured cells and alter phenotype, functional characteristics, and potency as compared to the same cell type residing in the body. This is of crucial relevance during translational development of cell-based therapies, at which multipotent mesenchymal stromal cell (MSC) play a major role. Here we use the example of MSC isolation and culture to illustrate critical aspects of cell culture

(continued)

J. Burk (✉)
Justus-Liebig-University Giessen, Equine Clinic (Surgery, Orthopedics), Giessen, Germany
e-mail: Janina.burk@vetmed.uni-giessen.de

© Springer Nature Switzerland AG 2021
C. Kasper et al. (eds.), *Basic Concepts on 3D Cell Culture*, Learning Materials in Biosciences,
https://doi.org/10.1007/978-3-030-66749-8_3

procedures on a theoretical as well as practical level. Finally, this chapter culminates in an unconventional cartoon style illustration of the cellular perspective. Imagine what life might be like in a cell culture dish.

3.1   Introduction to MSC in Bench to Bedside Research

Multipotent mesenchymal stromal cells, often also referred to as mesenchymal stem cells (MSC), represent the basis for many cellular therapies. MSC are adult progenitor cells that can be harvested by minimally invasive means from the postnatal organism, thus their clinical use can comprise autologous therapies, i.e. using cells from the patient, and is ethically well-accepted. MSC were initially promoted as a promising therapeutic tool due to their differentiation potential into mesenchymal lineage cells, aiming to replace damaged cells in the patient. However, over the following years, it became evident that the true potency of these cells lies in their modulatory and trophic actions [1], with the modulation of immune cells probably being the best characterized mode of action so far.

Therapeutic approaches using MSC include some applications which have already been translated into clinical use, e.g. for graft-versus-host disease [2], and many more applications which are still being researched in preclinical settings [1, 3]. Thus, many therapies are still at the early stages within the bench to bedside process described in Fig. 1.1. As already illustrated in this previous chapter, physiologically relevant cell culture models are crucial at this stage of translational research. If the in vitro model chosen is capable to reflect the critical aspects of the (patho)physiological environment in vivo, the results obtained will be more predictive for the treatment outcome in patients. Furthermore, with regard to manufacturing the putative MSC product, improved culture conditions during cell expansion help to maintain or trigger the potency of the cells before they are transplanted [4]. As a consequence, carefully chosen cell culture conditions will increase translational success, because they improve the physiological relevance of in vitro models and because they impact on MSC potency. Last but not least, the numbers of animals needed in preclinical studies could be reduced when adequate cell culture models are available.

As MSC represent a practically relevant example of primary cells used in basic research, with a clear focus on their therapeutic application, they will serve as an example throughout this chapter. We will illustrate the environmental changes during MSC cell isolation and expansion culture, including an unconventional and direct view from the cellular perspective.

3.2   MSC Isolation: Separation from the Niche

MSC have been isolated from a variety of tissues of the postnatal organism. Frequently used sources of MSC comprise bone marrow, cord blood, and adipose tissue. In tissues, MSC reside in quiescent state and in close proximity to small blood vessels, which, along with the expression of common marker antigens, has stimulated the hypothesis of their close relationship to pericytes [1, 5]. Besides these capillaries with their vascular endothelial cells and blood cells, tissues consist of specialized cells and extracellular matrix structures. Adipose tissue contains immune cells such as macrophages, fibroblasts, preadipocytes and, last but not least, adipocytes, all embedded in a collagenous extracellular matrix (Fig. 3.1, upper left). The obvious function of adipose tissue is to store fat as an energy reserve, but more recent research has also identified substantial endocrine activity in the tissue. Altogether, these different cell types, the extracellular matrix, and fluids with their soluble components contribute to the complex physiological MSC niche environment.

A typical standard protocol for MSC isolation from adipose tissue and subsequent MSC expansion is described below (see step-by-step protocol). We will first discuss some aspects of its theoretical background.

Adipose tissue is typically harvested from subcutaneous fat depots, with waste materials from plastic surgeries being a suitable source of human MSC for research purposes. Liposuction material can be used as well as tissue resected en bloc [6], which is illustrated here (Fig. 3.1, upper right). En bloc tissue resection does not disrupt the inner structures in the first place, although the tissue margins will be damaged on histological level. However, separation of the tissue from the circulation, together with the ongoing cell metabolism, entails successive changes of its physical and chemical inner milieu (for details, see Chap. 1.2.3), among others the accumulation of metabolites. In order to slow down cell metabolism, tissues are often cooled until they are further processed (Fig. 3.1, lower left).

Once transported to the laboratory, to isolate the MSC from adipose tissue (and also other collagenous tissues), traditional protocols rely on the disruption of the tissue structure. This is achieved by mincing the tissue, using scissors or scalpel, into small pieces, followed by tissue digestion using enzymes. Tissue digestion protocols may vary depending on tissue density, and may require different enzymes depending on the extracellular matrix composition. Frequently used enzymes are collagenases and dispases, the latter cleaving fibronectin but also some collagens. For adipose tissue digestion, crude clostridial collagenase is traditionally used [7] and sufficient to liberate the cells within a few hours when incubated at body temperature. During digestion, the enzymes attack the peptide bonds in collagens, thus destroying the extracellular matrix structure, resulting in a suspension of cells in the enzyme solution (Fig. 3.1, lower right). However, enzymatic cleavage is not entirely specific for collagen, and it should be noted that non-purified collagenase additionally contains a smaller amount of other enzymes that are also involved in tissue disruption [7]. This may result in the loss of surface antigens on the cells isolated, as observed after dispase digestion but also, to a smaller extent, after collagenase digestion

[8]. Although lost surface antigens partially recover within 24 h [8], this is likely to affect cellular functions.

3.3   Standard MSC Culture: Adaptation to an Artificial Environment

Following enzymatic tissue digestion, the tubes containing isolated cells, tissue remnants, and collagenase solution are centrifuged, which allows to separate the cell fraction including the MSC from fat and collagenase solution, based on their different density. The respective cell fraction (referred to as stromal vascular fraction) is then collected, washed in a buffer solution, and resuspended in cell culture medium for seeding. Note that this stromal vascular fraction still contains several different cell types, along with cellular debris. In numbers, depending on donor and harvest site, only 1–10% of these cells will correspond to MSC [5]. Hence, the MSC need to be selected and expanded during further processing. There are some options to actively sort cells and thereby select or deplete certain cell types, mostly based on immunological techniques using fluorochrome- or magnetically labeled antibodies binding to surface antigens present only on specific cell types. However, standard procedures in many laboratories do not involve cell sorting, but rely on the circumstance that MSC, similar to fibroblasts, are plastic-adherent and highly proliferative in traditional standard cell culture conditions. Thus, it is a typical approach to simply seed the stromal vascular cell fraction in standard plastic culture dishes (Fig. 3.2, upper left) and expand the cells for a few passages, which will increase the homogeneity of the cell population over time. Yet, irrespective of the benefit and simplicity of this approach, note that the MSC are subjected to a drastic change of environment at this step: They need to accomplish a transition from the niche environment, where they are embedded in soft 3D extracellular matrix and in close contact to different cell types, to a stiff 2D plastic surface with few possible cell-to-cell contacts.

Several more components of the artificial environment are equally associated with fundamental changes to the newly cultured cells. They include, e.g., oxygen partial pressure and nutrient supply (Chap. 1.3). For example, glucose is an important energy substrate and nutritional component of cell culture media. Basal media are available with different glucose content, typically either 1 g/L (low glucose) or 4.5 g/L (high glucose). Choosing a medium glucose concentration of 1 g/L will provide physiological conditions in this respect, as in vivo blood glucose levels below 140 mg/dL are considered as normal. However, keeping in mind that even if cells will proliferate well in high glucose medium, this practically represents diabetic conditions (Fig. 3.2, upper right), which in turn can affect MSC function [9]. While glucose concentrations represent an intuitive example of why culture medium composition may affect the cultured cells, it is known that media formulations generally impact on MSC properties and the maintenance of their potency [10].

After MSC seeding on plastic dishes, they will adhere, proliferate, and start to communicate again by releasing soluble factors and extracellular vesicles into the culture medium. Furthermore, MSC will synthesize a new extracellular matrix network, which forms a layer on the cell culture dish to which the cultured cells connect (Fig. 3.2, lower left). Thus in general, MSC are capable to adapt to new environments. On the one hand, this implies that to some extent, they may still be viable and proliferate even when culture conditions are suboptimal. Yet, this is not desirable as, on the other hand, alterations can be expected with respect to the MSC regenerative mode of action and potency, depending on the cell culture environment. Especially long-term MSC expansion eventually leads to loss of potency (Fig. 3.2, lower right) and genetic instability, the latter being a major issue regarding the safety of MSC therapies [11, 12]. Consequently, as described above, success of translating MSC-based therapies is hampered if in vitro models are inadequate and not predictive for the in vivo situation, and if cell expansion strategies are not suitable to maintain the desired phenotype and potency.

3.4   3D MSC Culture Approaches: Happy Ending Within Reach?

To overcome these limitations, there has been considerable research activity to improve cell culture conditions for MSC and develop strategies to triggering MSC potency during their in vitro expansion. Such so-called priming approaches include pro-inflammatory stimulation, hypoxic or 3D cultures, and many more [4]. Furthermore, in line with the demand for functional MSC characterization and tailored MSC potency assays, the numbers of published 3D and/or co-culture in vitro models are rising. One promising example is to combine 3D MSC culture in hydrogels with co-culturing endothelial cells [13, 14]. This co-culture in a 3D setting allows for communication of cells with their mutual neighbors and reflects important aspects of the MSC niche (Fig. 3.3). Essentials and procedures for 3D culture in hydrogels as well as for culturing endothelial cells are described in depth in Chap. 5 and Chap. 6, respectively.

These developments are overall highly encouraging. Nevertheless, it must not be neglected that the implementation of optimized culture conditions by different laboratories, as well as the use of specifically tailored in vitro models, is currently strongly hampering comparability of studies performed by different groups. The critical future step will be to identify the most suitable optimized culture conditions and in vitro cell culture models and to define those as standard conditions and reference procedures.

3.5   The Researcher's Versus the Cellular Perspective

After these theoretical considerations, you may ask yourself what MSC isolation and expansion will look like in practice. In the following, we will first provide a step-by-step protocol for MSC isolation and expansion, which illustrates the procedures performed in the laboratory—from the researcher's perspective.

MSC Isolation From Adipose Tissue Resected En Bloc
Note: The procedure presented here is a widely feasible standard approach, the shortcomings of which are discussed above. Thus, this protocol is not intended to serve as best practice example. It is mainly intended to document the researcher's perspective in the procedural steps corresponding to the situations experienced by the isolated cells as illustrated in the cartoon below (Figs. 3.1, 3.2, 3.3).

1. Use adipose tissue harvested from subcutaneous fat depots.
2. Store the tissue in a buffer solution at 4 °C until further processing, ideally not longer than overnight.
3. Transfer the vessel with the adipose tissue sample into the biosafety cabinet and work with sterile equipment.
4. Remove possible skin and fibrous tissue remnants. Mince the adipose tissue to pieces of approximately 1 mm in diameter using scalpel or scissors.
5. Transfer the minced tissue into 50 ml centrifuge tubes containing collagenase I working solution (approximately 1 g tissue wet weight per 10 ml). The working solution is prepared with a collagenase I (e.g., Gibco™, # 17100-017, ThermoFisher Scientific) concentration of 0.8 mg/mL, dissolved in Hank's buffered saline with calcium and magnesium.
6. Incubate at 37 °C under permanent shaking. If tissue was minced thoroughly, it should be digested within a maximum of 4 h.
7. Centrifuge at 400 xg for 5 min at room temperature.
8. Penetrate the fatty layer with a 5 ml serological pipet, harvest the cell pellet, and transfer it to a new 50 ml centrifuge tube.
9. Add phosphate buffered saline to a total volume of 50 ml.
10. Centrifuge at 400 xg for 5 min at room temperature.
11. Discard the supernatant and resuspend the cell pellet in 5 ml cell culture medium (e.g., high glucose Dulbecco's modified Eagle's Medium supplemented with 10% fetal bovine serum and 1% penicillin-streptomycin).
12. Count the cells using a hemocytometer.
13. Add the required volume of cell culture medium and seed the cells on tissue culture plastic dishes at a density of 20,000 cells per cm^2.
14. Incubate the cells in a humidified atmosphere with 5% CO_2 at 37 °C, for 1–2 days.
15. Discard the cell culture medium containing the non-adherent cells and replace with fresh medium.
16. Incubate the cells in a humidified atmosphere with 5% CO_2 at 37 °C until cultures reach 80% confluency, with a medium change twice weekly.
17. Passage cells at 80% confluency using 1x trypsin-EDTA after thorough washing, and seed cells at a density of 3000 per cm^2.
18. Expand the cells as described until the desired cell number is obtained.

This chapter would not be complete if we did not take a closer look at what is happening in the culture dish at the same time—from the cellular perspective. Have a look at the cartoons below (Figs. 3.1, 3.2, 3.3) and try to connect the images with the respective protocol steps as well as with the corresponding theoretical background!

Take Home Message

MSC are subject of intensive translational research activities. They may survive drastic changes of environment during in traditional vitro culture. However, this may still be associated with the loss of their original phenotype and potency, which hampers the translation process. Therefore, it remains crucial to aim for more physiologically relevant in vitro models and cell expansion strategies, where 3D cell cultures and co-cultures would be very promising approaches.

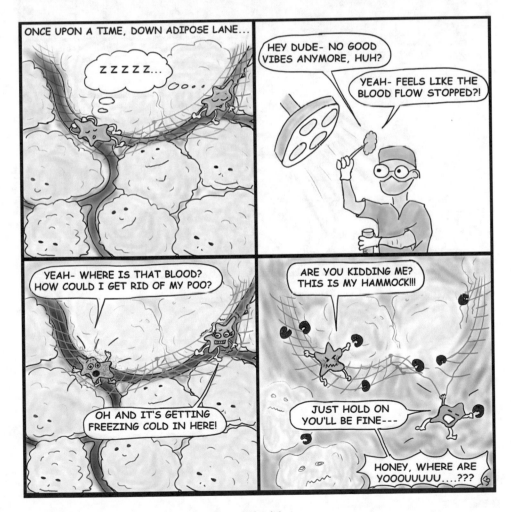

Fig. 3.1 MSC isolation and separation from their niche

Fig. 3.2 MSC culture and adaptation to an artificial environment

Fig. 3.3 A possible happy ending in 3D hydrogel co-culture

References

1. Andrzejewska A, Lukomska B, Janowski M. Concise review: Mesenchymal stem cells: from roots to boost. Stem Cells. 2019;37:855–64. https://doi.org/10.1002/stem.3016.
2. Galipeau J, Sensébé L. Mesenchymal stromal cells: clinical challenges and therapeutic opportunities. Cell Stem Cell. 2018;22:824–33. https://doi.org/10.1016/j.stem.2018.05.004.
3. Levy O, Kuai R, Siren EMJ, Bhere D, Milton Y, Nissar N, et al. Shattering barriers toward clinically meaningful MSC therapies. Sci Adv. 2020;6:eaba6884. https://doi.org/10.1126/sciadv.aba6884.
4. Noronha N d C, Mizukami A, Caliári-Oliveira C, Cominal JG, Rocha JLM, Covas DT, et al. Priming approaches to improve the efficacy of mesenchymal stromal cell-based therapies. Stem Cell Res Ther. 2019;10:131. https://doi.org/10.1186/s13287-019 1224-y.
5. Baer PC, Geiger H. Adipose-derived mesenchymal stromal/stem cells: tissue localization, characterization, and heterogeneity. Stem Cells Int. 2012;2012:812693. https://doi.org/10.1155/2012/812693.
6. Schneider S, Unger M, van Griensven M, Balmayor ER. Adipose-derived mesenchymal stem cells from liposuction and resected fat are feasible sources for regenerative medicine. Eur J Med Res. 2017;22:17. https://doi.org/10.1186/s40001-017-0258-9.
7. Williams S. Collagenase lot selection and purification for adipose tissue digestion. Cell Transplant. 1995;4:281–9. https://doi.org/10.1016/0963-6897(95)00006-J.
8. Autengruber A, Gereke M, Hansen G, Hennig C, Bruder D. Impact of enzymatic tissue disintegration on the level of surface molecule expression and immune cell function. Eur J Microbiol Immunol. 2012;2:112–20. https://doi.org/10.1556/EuJMI.2.2012.2.3.
9. Kornicka K, Houston J, Marycz K. Dysfunction of Mesenchymal stem cells isolated from metabolic syndrome and type 2 diabetic patients as result of oxidative stress and autophagy may limit their potential therapeutic use. Stem Cell Rev Rep. 2018;14:337–45. https://doi.org/10.1007/s12015-018-9809-x.

10. Yang Y-HK, Ogando CR, Wang See C, Chang T-Y, Barabino GA. Changes in phenotype and differentiation potential of human mesenchymal stem cells aging in vitro. Stem Cell Res Ther. 2018;9:131. https://doi.org/10.1186/s13287-018-0876-3.
11. Binato R, Fernandez S, de T, Lazzarotto-Silva C, Du Rocher B, Mencalha A, Pizzatti L, et al. Stability of human mesenchymal stem cells during in vitro culture: considerations for cell therapy. Cell Prolif. 2013;46:10–22. https://doi.org/10.1111/cpr.12002.
12. Neri S. Genetic stability of Mesenchymal stromal cells for regenerative medicine applications: a fundamental biosafety aspect. Int J Mol Sci. 2019;20(10):2406. https://doi.org/10.3390/ijms20102406.
13. Loibl M, Binder A, Herrmann M, Duttenhoefer F, Richards RG, Nerlich M, et al. Direct cell-cell contact between mesenchymal stem cells and endothelial progenitor cells induces a pericyte-like phenotype in vitro. Biomed Res Int. 2014;2014:395781. https://doi.org/10.1155/2014/395781.
14. Zhang X, Li J, Ye P, Gao G, Hubbell K, Cui X. Coculture of mesenchymal stem cells and endothelial cells enhances host tissue integration and epidermis maturation through AKT activation in gelatin methacryloyl hydrogel-based skin model. Acta Biomater. 2017;59:317–26. https://doi.org/10.1016/j.actbio.2017.07.001.

Biological, Natural, and Synthetic 3D Matrices

4

Viktor Korzhikov-Vlakh and Iliyana Pepelanova

Contents

What You Will Learn in This Chapter

The cultivation of mammalian cells in 3D is becoming increasingly important, as it becomes clear that cells display different morphology, signaling, gene and ultimately protein expression in comparison to conventional 2D cultivation. There is a wide variety of materials available for the 3D cultivation of cells, ranging from hydrogels that can be used to encapsulate cells, to porous 3D matrices, which can be seeded with cells. In this chapter, we will review the range of materials available for 3D cell

(continued)

V. Korzhikov-Vlakh
Institute of Chemistry, Saint-Petersburg State University, St. Petersburg, Russia

I. Pepelanova (✉)
Institute of Technical Chemistry, Leibniz University Hannover, Hannover, Germany
e-mail: pepelanova@iftc.uni-hannover.de

culture and will discuss their preparation, advantages, and limitations. We structure the discussion along the different ways in which cells are introduced into the matrix. Also, we lightly touch upon fabrication methods for creating porous matrices without going into a general discussion of fabrication methods. The chapter concludes with aspects to consider when selecting materials for 3D cell cultivation.

4.1   Introduction

In recent years, it has become increasingly clear that to obtain a more accurate understanding of cellular biology, it might be required to culture cells in surroundings, more closely related to their natural environment. Indeed, under physiological conditions, most cells grow within a complex three-dimensional (3D) environment, the fibrous, interconnected network of the extracellular matrix (ECM) which provides mechanical support, but also instructs cellular processes like adhesion, differentiation, gene expression and morphology [1]. It has now been shown by multiple studies that cells grown in a 3D substrate, as opposed to classic culturing in 2D, display different expression profiles [2, 3], show different behavior in drug testing studies [4, 5], and behave differently as an in vitro model (e.g., in cancer research) [6]. As a result, various methods have been developed for culturing cells in 3D, which can generally be differentiated into scaffold-free and scaffold-based techniques. In this chapter, we will turn our attention to the variety of materials used for 3D cell culture. For a discussion of scaffold-free cell aggregates and cell spheroids, readers are referred to Chap. 9 in this book ("Scaffold-free 3D cell culture").

Biologists, material scientists, and bioengineers have developed a range of 3D matrix materials and fabrication techniques, which allow researchers to mimic the natural extracellular microenvironment. Of course, the composition and structure of the ECM varies between different tissues in the organism, but understanding its general features and function guides the design of materials for 3D cultivation. There has been so far no consensus on the "ideal" 3D material or culture conditions to use for a specific cell type, but myriad new biomaterials have been used for in vitro 3D culture including metals, ceramics, composites, and polymers [7]. In this book chapter, we will focus mainly on polymers, because their design versatility has made them the most widespread materials for 3D culture.

The polymers used for 3D cell culture can be derived from natural materials or from synthetic derivatives [8]. Natural biopolymers are frequently structural proteins derived directly from the ECM like collagen, elastin, fibronectin, laminin, gelatin, Matrigel®, and fibrin. Other natural materials are not based on proteins, but on complex polysaccharides such as alginate, agarose, hyaluronic acid, chitosan, and chitin. Biopolymers have many advantages. For example, in the case of ECM-derived proteins, they already possess inherent bioactivity and cell-promoting properties. However, since they are derived from natural sources, biopolymers are frequently not precisely defined in their composition, leading to variability within batches and poor reproducibility between experiments. Materials derived from animal sources (such as most protein ECMs) also exhibit some risk of pathogenic or viral transmission. Finally, the mechanical properties of many natural materials are difficult to control and inadequate for long-term cellular cultivation. Synthetic polymers, in contrast, possess a well-defined composition and offer better control over mechanical characteristics. Examples of synthetic polymers used in 3D cell culture include poly(ethylene glycol) (PEG), poly(lactic acid) (PLA), poly(ε-caprolactone) (PCL), poly (vinyl alcohol) (PVA), poly(glycolic acid) (PGA), poly(lactic-co-glycolic acid) (PLGA), and many more. Frequently, material scientists combine the best of both polymer worlds and create semi-synthetic or (bio)hybrid materials. These contain some biological component responsible for cell-promoting properties, such as cell adhesion motifs, and a synthetic component, which imparts the structure with mechanical integrity [9].

Regardless of its source of origin, a material used for 3D cell culture should support cell growth and be highly biocompatible. Apart from this, it should possess an interconnected, porous structure, like the native ECM, which allows the diffusion of nutrients, metabolites, signaling molecules, and the transport of waste substances. Since the ECM represents a gel-like, highly hydrophilic structure, wettability is also an important criterion here. Moreover, the material should offer adequate mechanical properties, as mechanotransduction is known to influences cell phenotype [10]. Other material features are specific to the purpose of experiment and may include transparency for the purpose of microscope imaging, matrix degradability if the construct is to be implanted, and special cell-instructive properties like growth factor binding ligands for spatial and temporal control of tissue development.

The choice of a biomaterial in a certain experiment, as well as its functional design, always depends on the type of cell or cells to be studied, on the intended application and on the fabrication methods available in a specific laboratory. For instance, in vitro applications require an approximation of the cellular microenvironment, and this in a degree of complexity demanded by the experimental purpose. The material should be able to some extent to mimic the 3D organization and specialized function of the modeled tissue compartment. Typical applications in this area range from models for drug testing studies, which approach physiological conditions and can reduce the number of animal experiments [11], to tumor models and basic cell biology studies [12]. In the clinical field, 3D culture is used for the cultivation of stem cells, as well as in the development of regenerative therapies in tissue engineering. Here, the goal is to create a functioning implant that will assist or replace damaged tissue function in the patient. The requirements placed on the matrix are to support cell growth of course, but it should also slowly degrade in the body as cells re-establish their own matrix. This degradation should not cause allergic or adverse physiological reactions in the organism. Moreover, it must be possible to fabricate the implanted construct in a size and shape similar to the defect to be replaced [13].

There are a variety of fabrication techniques available to manipulate natural and synthetic polymers, from the macro- to the nanoscale, thus providing researchers with structural control which can mimic the hierarchical organizational complexity of living

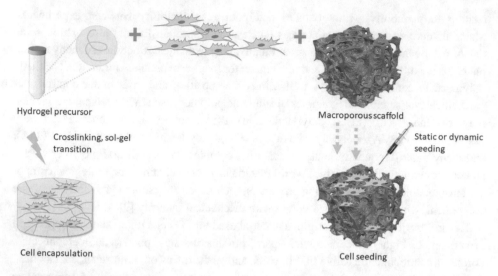

Fig. 4.1 3D cell cultivation is performed by cell entrapment in hydrogels, hydrophilic polymers which undergo crosslinking or sol–gel transition to form a solid construct. Alternatively, 3D cell culture is also carried out by seeding cells onto a pre-fabricated 3D scaffold which is then cultivated

tissue [14]. The choice of fabrication method influences the structural organization of the material, especially on the microscale. For example, electrospinning provides a tightly woven mesh network of polymers [15], while salt leaching can create pores of defined size and distribution within a macroporous structure [16]. And 3D printing can create constructs with highly compartmentalized and ordered architectures [17]. An in-depth review of the various biofabrication methods is beyond the scope of this chapter. Here we refer the reader to excellent reviews on this subject matter [13, 14]. We will venture into a limited discussion of fabrication methods as we consider the methods used to create 3D porous scaffolds from various natural and synthetic materials.

The chapter will proceed to present the most widely used polymers for 3D cell culture from natural and synthetic origin. We will review these materials especially under the aspect of whether cells are encapsulated or seeded on these materials (Fig. 4.1). Finally, we will round up with a discussion of considerations for material selection and a short outlook where material engineering for 3D cell cultivation is going next.

4.2   Methods of Preparation

There are two main ways to introduce cells into the 3D matrix. One way is to seed them directly on the construct—where cells will colonize the interconnected, porous 3D matrix by migration [18]. Different methods are used for this purpose, which can be roughly divided into static and dynamic techniques. Static methods involve surface seeding of cells or injection of cells into the scaffold. Dynamic methods include some form of movement of the cell suspension through the porous construct and may involve perfusion in a bioreactor/spinner flask, or the application of an external force like centrifugation or filtration. The cell seeding procedure should not be detrimental to cell viability and would ideally lead to even distribution of cells throughout the scaffold [19].

Another method of introducing cells to a 3D matrix involves performing an encapsulation of the cells, by embedding the cells in the liquid precursor material, which then undergoes crosslinking by physical or chemical methods to form a solid, 3D construct. The gelation reaction should be mild and non-toxic to cells. Hydrogels, hydrophilic polymers which absorb many times their own weight in water, are usually the polymers used for such encapsulation [20]. Since hydrogels are the most commonly used platform for 3D cell culture, they have received their own chapter in this book (Chap. 7 "Hydrogels for 3D cell culture"), which discusses hydrogels and methods for their analysis in detail. In this chapter, we discuss hydrogels from the perspective of their crosslinking chemistry, the mechanisms with which the polymer chains entrap cells into a 3D network.

4.2.1 Cell Entrapment by Encapsulation—Hydrogels

Hydrogels form hydrophilic networks of high water content, entrapping cells in a gel-like matrix, which bears close resemblance to most soft tissues (Fig. 4.2). All mass transport takes place as slow diffusion, creating gradients of signaling molecules and nutrients,

Fig. 4.2 3D cell cultivation in semi-synthetic GelMA hydrogels [21, 22]; Top: Cell encapsulation procedure by (**a**) mixing cell suspension with hydrogel precursor and pipetting in appropriate molds (**b**) crosslinking of hydrogel by UV light exposure and (**c**) polymerized bulk hydrogel; Bottom: hAD-MSCs encapsulated in GelMA hydrogels of different degree of functionalization (DoF) showing decreased cell spreading and network formation with increased crosslink density, Calcein-AM staining, scale bar: 1000 μm

similar to the situation in the native ECM [23]. Hydrogels can undergo physical crosslinking based on non-covalent bonding such as hydrogen bonding, dipolar interactions, and Van der Waals forces. Alternatively, hydrogels could be formed by chemical crosslinking, which involves the creation of covalent bonds. In general, the material stiffness and to some extent the surface topography of hydrogels can be influenced by adjusting the concentration of precursor and/or crosslinker [24]. We will now shortly review some of the most common principles of hydrogel formation for the encapsulation of cells in perspective of whether the interactions occur naturally, as found within the toolkit of natural processes, or by synthetic crosslinks, by using the ingenious methods of chemists and material scientists.

4.2.1.1 The Way of Nature

Many of the most widely used hydrogels for 3D cell culture are widely available natural polymers, which undergo a sol–gel transition through a natural process of fibril self-association, hydrophobic, ionic interactions, or enzymatic crosslinking. The liquid precursors can be obtained commercially from a variety of chemical vendors [25]. These are then mixed with cells and the appropriate trigger of gelation is introduced to the system (e.g., pH shift, ions, enzyme addition), leading to formation of a solid, 3D network of encapsulated cells.

Natural protein materials like collagen I, gelatin, or basement membrane extract (Matrigel®) are very popular as 3D cell culture platforms. As they are derived from the ECM, such protein hydrogels naturally possess cell adhesive and cell-signaling sequences which promote cell growth, proliferation, and differentiation [26]. Cells can also degrade the proteins via cellular proteases, remodeling the network as they grow. Protein hydrogels consist of solubilized fibers. If sold and formulated in the liquid state, the precursor is kept at low temperature and/or pH to inhibit gelation. In the laboratory, the precursor is mixed with cells at an appropriate concentration, and then temperature and pH are raised to physiological conditions. This leads to a self-association of the individual protein fibrils, gradually leading to the formation of a solid network. Encapsulation is usually complete within 30 min. Because the mechanism of network formation is physical self-association, all protein hydrogels tend to form weak gels with poor mechanical properties. This is an important drawback of such materials, and they are frequently reinforced by adding covalent crosslinks by using synthetic or enzymatic methods (see also synthetic tricks of chemists).

The natural self-assembly of protein structures has inspired protein engineers to design synthetic peptide hydrogels. These contain amphiphilic or other complementary sequences, which can assemble into higher macromolecular nanostructures such as beta-sheets [27]. Peptide hydrogels are also commercially available (e.g., PuraMatrix™, a 16-residue peptide composed of a repeating sequence of arginine-alanine-aspartate-alanine) [28]. Such materials have tremendous potential because they are effective mimics of the ECM, while allowing complete control of adhesion sites, degradation, and organizational hierarchy of the hydrogel [29]. Disadvantages of peptide hydrogels are their relatively high cost, which prohibits their use in large-scale cultivations. In addition, in a manner similar to natural proteins, the mechanical properties and long-term cultivation stability of synthetic peptide hydrogels is rather low, due to the weak non-covalent intermolecular interactions responsible for macrostructure assembly.

Other natural hydrogels assemble into solid structures through ionic interactions. The most famous and widely used example is alginate, a polysaccharide of brown algae. Alginate is a linear block copolymer of β-D-mannuronic acid and α-L-guluronic acid units. Cells can be easily immobilized in alginate by mixing with precursor and adding divalent cations to the construct (e.g., calcium, magnesium, barium). The cations form ionic bridges with the alginate chains, stabilizing the macromolecular structure. Cell recovery is possible by removal of the ions with chelating agents. Alginate is transparent and non-toxic, however, it must be modified with adhesive sequences if cells are to attach and display long-term growth on the material. Also, the long-term cultivation stability of alginate is compromised by diffusion of its calcium into the surrounding medium, leading to degradation [30]. Other example of hydrogels formed upon electrostatic interactions involves the use of two precursors of oppositely charged biopolymers. Examples include hydrogels of chitosan (positively charged amino groups) and hyaluronic acid/alginate (negatively charged carboxylic groups). The formation of such polyplexes enhances the mechanical properties of the individual hydrogels [31].

Hydrogels can be assembled with covalent crosslinks by the action of enzymes. Since enzymatic reactions take place at physiological conditions, this makes them very attractive options for introducing crosslinks to polymers in the presence of cells. A prominent example used in 3D cell culture is the polymerization of the blood protein fibrinogen to fibrin via the enzymatic action of thrombin. Fibrin possesses multiple binding sites for various cell types and growth factors, which makes it an attractive material for cell cultivation [32]. For use in 3D cell culture, fibrinogen can be isolated from blood (human or bovine), or can be purchased as a clinical tissue sealant (e.g., Evicel® by Johnson & Johnson or Tisseel® by Baxter). The rapid reaction with thrombin polymerizes fibrinogen with embedded cells into a fibrin matrix. However, as with other natural materials, there are drawbacks associated with batch-to-batch variation and with rapid proteolytic degradation, especially on long-term cultivation [33]. For these reasons, 3D fibrin cultures are frequently supplemented with aprotinin, a protease inhibitor [34].

Other enzymes are also used for crosslinking hydrogels with the goal of encapsulating cells (Fig. 4.3). The most established examples in 3D cell culture include transglutaminase and horseradish peroxidase [35]. Transglutaminases are a class of enzymes, which are naturally involved in many physiological cell–matrix interactions. These enzymes are also called "tissue or biological glue" and they catalyze the reaction between a free amine group and the γ-carboxamide group of a protein to create an amino bond. Co-factors are not required, and the reaction is complete within 20 min. A variety of hydrogels including collagen [36] and gelatin [37], but also peptide-PEG conjugates [38] have been linked by transglutaminase into stable networks with cells.

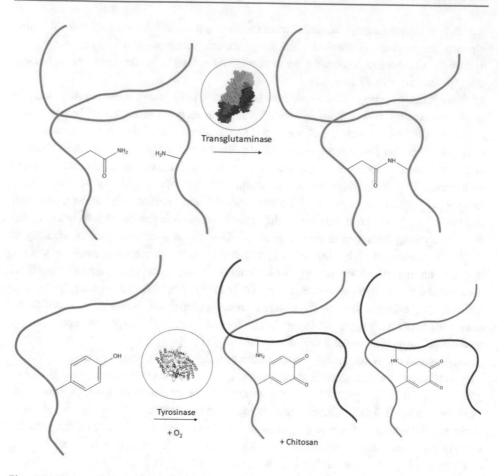

Fig. 4.3 Enzymatic reactions used for crosslinking hydrogels; Top: the enzyme transglutaminase catalyzes the formation of isopeptide bonds between γ-carboxamide groups of, e.g., glutamine residues and ε-amino groups of, e.g., lysine residues. The resulting bonds are resistant to proteolytic degradation. No co-factors are required. Bottom: the enzyme tyrosinase requires oxygen to oxidize phenols (like tyrosine or dopamine). The activated quinones proceed to react with amino or hydroxyl groups in a Michael-type addition reaction

Horseradish peroxidase (HRP) catalyzes the oxidative coupling of phenol-rich polymers (i.e., those containing phenol and aniline derivatives) in the presence of hydrogen peroxide. The crosslinking is completed within seconds to minutes and the resulting hydrogel properties can be controlled by adjusting the proportions of reactants. Some natural materials like gelatin, silk fibroin, and pectin can be directly crosslinked with HRP, because of their high content of native tyrosine, phenylalanine, and feruloyl groups, respectively. Most other natural and synthetic polymers have to be previously modified with tyramine or hydroxyphenylpropionic acid, to create a phenol-rich polymer for the formation of hydrogels by HRP [39]. Especially tyramine-conjugates of polysaccharide-based polymers like dextran [40] and hyaluronic acid [41] have a proven track record as valuable matrices for 3D cell culture.

4.2.1.2 Synthetic Tricks of Chemists

Synthetic tricks leading to the formation of hydrogels capable of cells entrapment include application of synthetic thermosensitive polymers and chemical crosslinking.

Some synthetic polymers possess the so-called low critical solution temperatures (LCST). When below this temperature, water will dissolve such polymers to form a liquid solution, in which cells can be suspended. Nevertheless, when the solution is heated above the LCST, the polymer undergoes gelation and entraps the cells. The mechanism of gelation is based on the fact that water molecules do not really "like" to be bound to such polymers, but at room temperature water does not have enough energy to escape from such interactions. However, upon heating water gets enough energy and quits the polymer coil, leading to its collapse. The interested reader can find a more detailed description of this phenomena in this publication [42]. Typical examples of such polymers are poly (N-isopropylacrylamide) (PNIPAAm) [43] and poly(oligo(ethyleneglycol)methacrylate) (POEGMA) [44]. Both of these polymers can undergo gelling at physiological temperatures: 31–35 °C. Such polymers can be used on their own, but in order to provide better biocompatibility and interactions with cells there are combined with polysaccharides [45] or proteins [46]. Both PNIPAAm and POEGMA are non-degradable. If the application requires a biodegradable thermosensitive polymer, synthetic block-copolymers of polyesters, such as poly(ε-caprolactone-block-ethylene glycol-block-ε-caprolactone) (PCL-PEG-PCL) [47], as well as thermosensitive polypeptides [48], can be applied.

The covalent crosslinking of macromolecules to form hydrogels is another attractive method of cell encapsulation. There are two words characterizing this approach, which make this method so alluring for encapsulation, and they are—versatility and controllability. The first one means that chemical groups, which cause crosslinking and subsequent gelling, could be introduced in practically any synthetic or natural polymer of interest. The second word implies that the number of introduced chemical groups determines the number of crosslinks formed, which are crucial for control of gel mechanical properties, permeability, and stability. Chemical bonds could be irreversible or hydrolytically/enzymatically degradable. In this sense, the chemistry of crosslinking should be chosen in accordance with the proposed material fate in a long-term perspective.

The chemical reactions involved in cell encapsulation have some limitations dictated by biosafety restrictions. Most methods of cell encapsulation begin with dispersion of cells in the aqueous solution of hydrophilic macromolecules (hydrogel precursor) and proceed with the subsequent formation of crosslinks. Thus, the crosslinking chemical reactions are intended to proceed in water solutions at mild, nearly physiological conditions: temperature 20–37 °C, neutral pH, etc. The reactions should not be sensitive to buffer salts, which are usually present in the system. Obviously, the reaction should neither involve nor provoke the formation of toxic chemical substances [49]. Natural, synthetic, and semi-synthetic macromolecules can be used for the formation of cell-laden hydrogels. The chemical

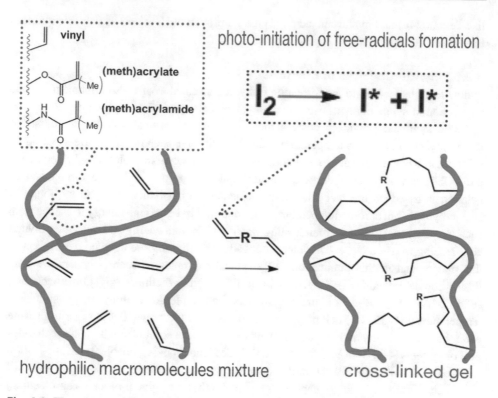

Fig. 4.4 The scheme of photo-initiated free radical crosslinking of hydrophilic natural or synthetic polymers bearing double bonds leading to gel formation

moieties of natural polymers usually do not permit chemical reactions at the above-mentioned conditions. Thus, the encapsulation of cells by chemical crosslinking requires the application of synthetic polymers or semi-synthetic (hybrid) derivatives of natural ones. Hybrid polymers are natural polymers with introduced chemical groups allowing the conduction of specific and efficient crosslinking chemical reactions.

One of the classical chemical routes for crosslinking of hydrophilic macromolecules into hydrogels is the free radical reaction involving unsaturated double bonds—vinyl or (meth)acrylate groups (Fig. 4.4).

Such moieties can be introduced in both synthetic and natural polymers resulting in the formation of the so-called macromonomers. A macromonomer is a macromolecule, which can participate in a free radical polymerization reaction in a similar manner to low molecular weight monomers due to the presence of double bonds in its structure. To formation of crosslinks usually requires the addition of two low molecular compounds: crosslinker with two double bonds in one molecule and initiator. The decomposition of initiator forms free radicals, which initiate the crosslinking. This could be carried out in mild conditions by one of the following methods: (1) reduction–oxidation (Red-Ox) and (2) photo-initiated initiation. The Red-Ox initiation involves the application of a special initiator—ammonium persulfate (APS), which dissociates with formation of two sulfate radicals. The dissociation could occur at room temperature or even at temperatures below 0 °C, when it is catalyzed by tetramethylethylenediamine (TEMED). In fact, this initiation method is widely used for preparation of polyacrylamide gels for protein electrophoresis (SDS-PAGE). In several studies, the APS/TEMED initiation system was applied to encapsulate cells in thermosensitive PNIPAAm hydrogel crosslinked by N,N'-methylene-bis-acrylamide [50], as well as in hydrogels based on (meth)acrylated PEG, namely polyethylene-glycol dimethacrylate (PEGDMA) [51] and poly(trimethylene carbonate)-b-poly(ethylene glycol)-b-poly(trimethylenecarbonate)-acrylate (PEG-(PTMC-A)$_2$) [52]. In the latter study, the PEG-(PTMC-A)$_2$ was used to form crosslinks between methacrylated chondroitin sulfate macromolecules. However, it should be noted that the APS/TEMED initiating system could cause some toxicity for the encapsulated cells [51, 53] and can be recommended only in the cases when application of photo-initiated polymerization (see below) is not possible.

Photo-initiated polymerization is a cell-friendly approach to form covalent crosslinks via a free radical process. The realization of this method requires a special photoinitiator, which could form radicals under ultraviolet irradiation (e.g., 2,2-dimethoxy-2-phenylacetophenone, Irgacure 651) or when exposed to visible light (e.g., Camphorquinone). Such initiators could provoke cytotoxicity of the gel, but usually the cell viability is not affected by the very small concentrations of initiators used for photo-initiation [54]. This method is advantageous in many cases because it is fast and allows both spatial and temporal control. Spatial control means that polymerization could be initiated only in designated areas via application of light beam [55], laser beam [56], or photopatterning with the application of a mask [57]. In the latter case, the mask allows the light to pass through only at the desired regions.

As already mentioned, both synthetic and natural polymers can be modified to have double bonds in their structure. However, synthetic polymers usually do not provide good cell–material interactions. For this reason, (meth)acrylated natural polymers are usually applied as macromonomers, while synthetic polymers or oligomers serve as crosslinkers. The typical examples of such gels are those obtained by crosslinking of methacrylated (MA) gelatin (Gel-MA) [58], hyaluronic acid (HA-MA) [55], heparin (Hep-MA) [59], cyclodextrin (CD-MA) [60] by polyethylene-glycol di(meth)acrylate (PEGDMA). Nevertheless, in some cases methacrylated natural polymers are gelled to encapsulate cells without addition of crosslinker: GelMA [61], alginate-MA [62], GelMA mixtures with HA-MA [63], chitosan-MA [64], galactose [65], etc. Also synthetic polymers like PEGDMA can be used alone [54] or introduced into complex graft-copolymers, like chitosan-g-polyethylene glycol-g-methacrylate [66]. In the latter case, the obtained polymer can undergo both thermogelling and photocuring.

In addition, hydrophilic polymers can be covalently crosslinked and gelled by interaction of highly reactive chemical groups. Reaction of aldehydes with amino groups, resulting in formation of Schiff bases, is a quite fast reaction, which does not require any catalyst and produces water as the only by-product. For example, hyaluronic acid

(HA) oxidized with sodium periodate (NaIO$_4$) to give HA-aldehyde was allowed to interact with the amino groups of chitosan [67], resulting in encapsulation of epithelial cells. Another study [68] realized interaction of aldehyde-bearing cellulose nanocrystals with the amino groups of gelatin to obtain gels with well-defined compositional gradients in the desired direction. However, the presence of polyaldehyde could cause toxicity for the cells, because the aldehyde groups can easily react with primary amino groups of proteins on the surface of cells [69, 70].

The development of organic chemistry has led to the discovery of the so-called click chemistry reactions. These reactions proceed fast and quantitatively at physiological conditions and usually do not involve any initiators or catalysts. Moreover, they are very region-selective, which means that they do not affect other chemical moieties within hydrophilic polymers and cells. However, such reactions require special chemical groups to be introduced into the structure of the interacting macromolecules. There are two general types of "click" chemistry reactions, which are useful for cell encapsulation, namely, thiol-ene Michael addition and copper-free azide-alkyne cycloaddition (Fig. 4.5).

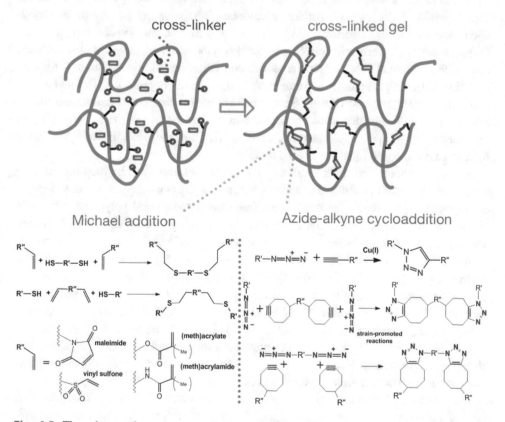

Fig. 4.5 The scheme of covalent crosslinking of hydrophilic natural or synthetic polymers via "click" chemistry approaches

The thiol-ene reaction represents a Michael-type reaction between thiol group (-SH) and α,β-unsaturated carbonyls (Michael acceptor) which result in formation of thioester according to anti-Markovnikov addition. Maleimides, vinyl sulfones, (meth)acrylates, and acrylamides can be used as Michael acceptors. This reaction allows encapsulation of cells via self-gelling of such components as gelatin-cysteine and PEGDA [71], glycidyl methacrylate derivatized dextran (Dex-GMA) and dithiothreitol (DTT) [72], gellan gum modified by divinyl sulfone and thiolated adhesive peptides [73]. In addition, Michael addition can be performed as a photo-initiated process. In this case, it possesses all advantages of photo-initiated polymerization, namely, spatial and temporal control of the gelling process. In this work [74], a special photo-responsive gelling macromolecule was synthesized, which produced -SH under exposure to visible light. The formed free thiols were crosslinked with four-arm PEG tetra-maleimide or a dextran maleimide. This approach allowed spatial and temporal control over gel properties, as well as resulted in highly biocompatible systems. Photo-curable gelling was also shown for gelatin-MA and gelatin modified by norbornene, with application of DTT or four-armed thiol-terminated PEG as crosslinkers [75].

Interestingly, that thiol-thiol reaction could be used for self-gelling without the addition of alkene. Bian with coauthors [76] have obtained hyaluronic acid with thiol groups (HA-SH), which was able to self-crosslink under physiological conditions. The only prerequisite for such a reaction is the treatment of HA-SH with dithiothreitol to destroy S-S bonds, which are immanently formed by oxidation of -SH groups with oxygen from the air.

During the last two decades, one can observe the great popularity of cycloaddition "click" reactions. Among them, 1,3-dipolar cycloaddition, which takes place between an azide and an alkyne, and results in the formation of 1,2,3-triazole, is of highest interest for bioconjugation. Both reactive groups can hardly be found within the molecules on the cells' surface (compare this with -SH and $-NH_2$ groups from the above-discussed reactions). Thus the reaction will not affect cell biochemistry and can be referred to as the so-called bioorthogonal chemistry [77]. The azide-alkyne reaction was firstly discovered to proceed under copper (I) catalysis, which could be toxic for cells. However, it was further revealed, that highly strained cyclooctynes react readily with azides to form triazoles under physiological conditions, without the need for catalyst addition [78]. The azide-alkyne "click" reaction was applied to encapsulate cells into gels formed by the reaction of hyaluronic acid, modified with cyclooctyne, with PEG-azide [79], or via interaction of chitosan-azide with PEG-alkyne [80]. Also the method could be applied for gelatin crosslinking [81]. The only current disadvantage of this method is that the polymers for reactions are poorly commercially available, while their synthesis in the laboratory requires quite complex synthetic procedures.

It should be noted that the area of biorthogonal crosslinking reactions for cell encapsulation is a developing scientific area, which includes research on Diels–Alder [82], nitrile oxide–norbornene cycloadditions, Staudinger ligation, and other perspective chemistries [77]. In this context, we would like to mention a recent publication in which a bioink based on poly(2-ethyl-2-oxazine) Diels–Alder reaction was shown to be useful for encapsulation of cells via melt-electrowriting [83].

4.2.2 Obtaining 3D Porous Matrices

The materials for the 3D culturing of cells require a porous structure in order to allow cell spreading and penetration inside the material. Moreover, porosity is important for the mass transport of nutrients and wastes. Generally speaking, the successful 3D culture of cells requires pores with diameter not <20 µm.

Other important characteristics, which are related to pores, are surface area and pore structure. The large surface area is important for cell attachment and spreading. However, the specific surface area is inversely proportional to the size of pores. In this respect, the formation of micropores in the walls of macropores could be optimal (Fig. 4.6). The uniformity of pores and pores interconnectivity are of importance, because they allow homogeneous cells distribution and migration within the material.

Taking into account the above-mentioned statements, the importance of methods to form 3D supermacroporous materials for successful 3D cultivation is obvious. It should be noted that the method of 3D porous support formation is usually highly dependent on the polymer material used for the matrix preparation.

4.2.2.1 Re-Housing in Evicted Homes: Decellularized Materials
In many respects, the attempt to reproduce a cellular niche by using a single type of (bio)polymer or even a simple mixture of polymers falls short of the true complexity of the

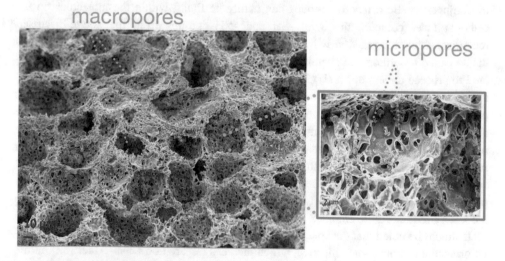

Fig. 4.6 PLA-based foam possessing both supermacropores and micropores in its walls

ECM. Therefore, many researchers have attempted to directly use the genuine ECM environment for the 3D cultivation of cells by using decellularized ECM substrates. The process of decellularization involves the removal of cells from a sample of human or animal tissue or organ. There are various methods to achieve this; the most commonly used techniques involve repeated freeze–thaw cycles, as well as enzymatic and chemical protocols [84]. The procedure should remove all native cells, as well as any xenogenic or allogenic components from the treated tissues, while attempting to preserve as much as possible of the ECM proteins and tissue-specific architecture [85]. The resulting decellularized ECM scaffolds exhibit great variety because they are all derived from different tissues and species, as well as being the product of various methods of decellularization. Interested readers are referred to the review of Song and Ott [86], describing the 3D cell culture in decellularized cartilage, liver, heart, bladder, bone, skin, and lung ECM. Indeed, decellularized ECM scaffolds provide an optimal environment for a particular cellular niche in terms of mechanical structure and architecture. However, they also exhibit numerous disadvantages. Due to their sourcing, decellularized scaffolds are extremely difficult to standardize. Their structural and compositional complexity makes them difficult to control and reproduce between different experiments. Moreover, the availability of ex-vivo tissues is not always granted and the decellularization procedure requires specific skills and great care.

4.2.2.2 Phase Separation for Obtaining Pores

There are many methods, which were developed in order to obtain porous materials applicable for 3D cell growth support: melt-molding, fiber bonding, porogen leaching, gas foaming, and phase separation [87]. Average pore size of more than 50 μm, pores uniformity, and interconnectivity are the main goals for these methods. In contrast to other techniques, the methods based on phase separation provide the highest versatility in terms of pore size and material morphology, and do not require complex equipment. By the way, the phase separation can be combined with other porous matrices fabrication techniques and this usually benefits matrix morphology [88].

   The foundation for such methods is the so-called thermally-induced phase separation (TIPS)—the thermodynamic technique that involves the separation of solvent and polymer phases in polymer solution due to physical incompatibility at certain temperature [89]. Such incompatibility leads to the separation of the homogeneous polymer solution into (1) polymer-rich and (2) polymer-poor phases. After removal of the solvent, the first phase forms the matrix, while the second one constitutes the pore space. The removal of solvent can be performed by two general methods: (a) leaching of the solvent by a non-solvent with its subsequent evaporation or sublimation or (b) sublimation of the frozen solvent. The example of case (a) is the preparation of poly(L-lactic acid) (PLLA) porous scaffolds by TIPS, after which the frozen tetrahydrofuran (THF, solvent for PLLA) could be washed out by non-frozen ethanol (non-solvent for PLLA) [90].

   It should be noted that there are two different TIPS cases, which result in different material pore size and morphology, namely solid–liquid phase separation (S-L PS) and liquid–liquid phase separation (L-L PS). S-L PS takes place when the interaction between polymer and solvent is quite strong (good solvent). Upon decreasing of the temperature, the crystallization of the solvent starts earlier than polymer precipitation due to phase separation. The growing crystals of the solvent displace the polymer into the polymer-rich phase. After removal of those crystals by one of the methods mentioned above, the porous material is formed. The size and form of the pores are dictated by the size and form of the solvent crystals. For example, if the freezing occurs in the glass, which is placed on the frozen surface of a refrigerator, the crystals will grow in the direction from the surface. After removal of the frozen solvent by one of the mentioned methods, anisotropic channel-like pores will be formed. The S-L PS is the basis of the so-called cryogelation and freeze-drying techniques (see below).

The L-L PS is observed when polymer–solvent interactions are weak. At a certain temperature of metastable condition, which is between binodal and spinodal on the phase diagram of the system, the separation of the visually homogeneous solution into polymer-rich and polymer-lean phases occurs. Notably, no polymer precipitation is observed at this metastable stage, but this could start if someone will cool down the system with one more degree. This metastable region is characterized by the coexistence of the polymer-rich phase and the solvent-rich phases. After attaining these conditions, the system should be quenched at the temperature, at which no crystallization will occur. Usually this is performed by treatment of the system with liquid nitrogen. After that the solvent should be removed by one of the methods discussed above [90].

The TIPS is a very versatile technique, because it places no limitations on the polymer to be used. Every polymer that can be dissolved in some solvent can be subjected to S-L or L-L PS, depending on the thermodynamics of polymer–solvent interactions in the system. Moreover, the S-L PS could be pushed to L-L PS by change of the solvent quality. For example, PLA undergoes S-L PS in pure dioxane, but when 10 wt% of water is added to dioxane, L-L TIPS is observed [91].

L-L-PS is the method of choice for preparation of matrices based on aliphatic polyesters (PLA, PCL, etc.), which could be successfully used for 3D cell cultivation [92]. It was shown by Peter X. Ma and colleagues, that depending on the conditions of TIPS, matrices with different morphologies could be obtained [93, 94]: at a high gelation temperature, a platelet-like structure was formed, while at a low gelation temperature, the formation of nano-fibrous morphology was observed.

If one wants to obtain a supermacroporous matrix for 3D cell culture based on hydrophilic biopolymers, such as chitosan [95], alginate [96], gelatin [97], or gelatin/hyaluronic acid [98], the cryogelation technique based on S-L PS could be applied with success. The formation of cryogels includes freezing of the solution or preformed gel, crosslinking interactions between macromolecules and removal of the solvent [99] (Fig. 4.7). As the solution is frozen, the growth of the solvent crystals takes place until they get in contact with other crystals. Thus, the solvent crystals form a common extended structure. The phase containing the polymer is concentrated around the crystals and is called unfrozen liquid microphase (ULMP). In fact, crystals of water are pushing the

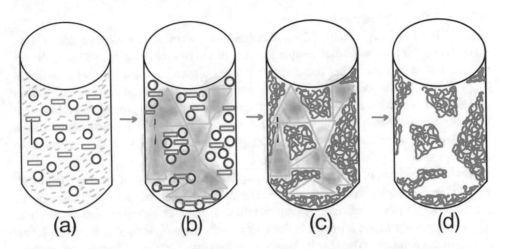

Fig. 4.7 Mechanism of cryogel formation, (a) gel precursor solution in water, (b) freezing and cryo-concentration, (c) crosslinking (formation of continuous phase), and (d) water crystals removal (porous matrix formation)

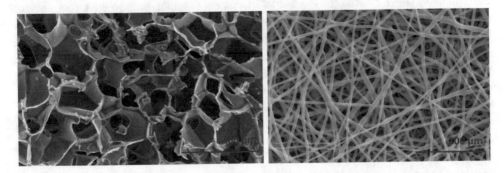

Fig. 4.8 Representative SEM images of pre-fabricated porous matrices for 3D cell culture obtained by different processing techniques; Left: the sponge-like microporous structure of a cryogel from alginate methacrylate crosslinked by PEG-dimethacrylate; Right: a fibrous mesh of a (90:10) PCL-chitosan blend engineered by electrospinning (SEM image courtesy of the Glasmacher group [101, 102])

polymer molecules together, which is called "cryo-concentration." At such concentrated conditions different chemical crosslinking reactions, such as cryo-polymerization [100], can run very efficiently even at temperatures below 0 °C. The last stage of cryogel formation is removal of the solvent crystals via drying or thawing, revealing the supermacroporous structure, which is useful for cells seeding and proliferation (Fig. 4.8, left).

4.2.2.3 Spinning Threads

It is well-known that natural ECM of connective tissues is fibrous. For this reason, materials which possess fibrous morphology are of interest for 3D cell cultivation. It was already discussed above, that phase separation techniques can produce fibers similar to ECM. However, it is sometimes difficult to control fiber alignment and diameters. Furthermore, phase separation for fibers formation represents a multistep procedure. At the same time, there is a one-step technique, which can rapidly produce fibrous materials with high control of the fiber diameter and material geometry. The name of this technique is electrospinning.

The implementation of electrospinning implies the availability of special equipment, which is neither expensive nor complex. Four main components of those are syringe pump, grounded target (fibers collector), spinneret (nozzle), and high-voltage power supply. The desired polymer solution is placed in the syringe and the pump is used to form droplet on the tip of the spinneret. Then the high-voltage is applied to this tip in order to generate an electric field and, consequently, charge on the droplet. Electrostatic repulsion provoked by this charge causes a force, which is opposite to the surface tension of the drop. When the intensity of electrical field overcomes the surface tension, a polymeric jet is ejected towards the grounded target. During the ejected jet movement from the drop to the target, solvent evaporation occurs and electrostatic forces cause the stretching of the polymer into thin fibers, which are collected as a 3D material. The geometrical form of the material can be controlled by shape, size, and rotation of the target collector, as well as by application of a single-, dual-, or coaxial nozzle. The fiber thickness and alignment is controlled by processing parameters (flow rate, voltage) or polymer solution properties (polymer molecular weight and concentration, viscosity, solvent) [103].

Electrospinning has found numerous applications in the formation of supports for 3D cell culture, because it allows for the rapid production of spatial and surface structures based on a broad range of synthetic and natural polymers and their combinations. The obtained matrices can mimic various natural tissues and structures [104]. The dominant number of works in this area are devoted to the formation and application of fibrous matrices based on polyesters (PCL, PLA) per se [105], as well as based on mixtures of polyesters with various natural polymers, such as PLGA-chitosan/PLGA-gelatin [106]. In the latter case, a mixed solution of PLGA, gelatin, and chitosan in 1,1,1,3,3,3-hexafluoro-2-propanol/acetic acid was prepared and used for electrospinning, resulting in the formation of three-layered biocomposite matrices. In some studies the combination of a relatively hydrophobic polyester with a relatively hydrophilic natural polymer was performed via blend formation [107], while in other the formation of polyester-biopolymer core–shell fibers was observed. In one of such studies the core was formed by PCL, while the biologically active shell was composed of collagen [108]. The PCL-collagen core–shell matrices provided better fibroblast adhesion than in the case of PCL alone, due to improved biointeractions.

The polyester itself is usually a bad substrate for cells adhesion. For this reason, the electrospun fibrous material based on PCL was subjected to oxidizing plasma treatment.

This treatment makes the surface of fibers more hydrophilic and increases cell adhesion. Furthermore, the obtained matrix was covered by intervertebral disk cells (IVD) dispersed in an alginate gel. The formed bi-compartmental structure allowed the co-culture of two cell types and resulted in increased formation of glycosaminoglycans and discogenic markers [109].

It is important to note that electrospinning of polyesters results in materials, which are insoluble in water. However, in the case of polyester combinations with biopolymers, efforts are needed to prevent the biopolymers from leaking from the structure. Sometimes crosslinking is performed to this purpose. For example, the stabilization of chitosan and gelatin within composite fibrous material could be performed by treatment with glutaraldehyde, which crosslinks the amino groups in the biopolymers to give Schiff bases [106]. In order to keep the alginate with dispersed cells integrated with the fibrous PCL matrix, the alginate was crosslinked by Ca^{2+} ions [109].

Post-electrospinning crosslinking could also be an issue in the case of preparing of electrospun matrices based only on hydrophilic biopolymers. For example, the alginate/gelatin matrices for 3D cell culture were obtained by application of calcium ions to crosslink alginate and gelatin-methacrylate to perform post-electrospinning photo-induced polymerization crosslinking [110]. However, the formation of chitosan-based fibrous materials was reported to proceed without any crosslinking [111]. Possibly the good hydrogen bonding and the poor solubility of chitosan in media with neutral pH are the reasons for the stability of such materials.

In conclusion, it is important to say that despite the long history of the electrospinning method, it keeps constantly developing. Some recent studies have reported the application of electrospinning for the encapsulation of cells inside core–shell capsules based on gelatin/alginate [112] and electrospinning with cells suspensions in a mixture of fibrinogen with polyethylene oxide [113]. Also, hybrid cell-laden alginate fibers reinforced with PCL were obtained and showed good MG63 cells viability and improved mechanical properties [114]. Similarly, muscle cells were successfully encapsulated into alginate-polyethylene oxide fibers [115]. Thus, electrospinning represents a type of technique that allows both scaffold material formation and cell encapsulation at the same time.

4.3   Aspects to Consider and Outlook

As this chapter has shown, there is a plethora of materials available for 3D cell culture and they all have their own associated advantages and disadvantages. The researcher should not be dazed by this wide range of choices, but carefully consider the purpose of experiment. How much of the material is available, can precursors be commercially obtained, to what extent are chemical skills required for functionalization, what fabrication methods are available in the laboratory, how essential are uniform batch-to-batch material properties in perspective of potential clinical studies or commercialization? For a start, however, the most important properties to consider in a 3D cell culture material is stability for the duration of experimental culture conditions, the ability of cells to adhere to the material, as well as adequate mechanical properties.

In addition, analytical aspects are also central to experimental planning and their importance should not be underestimated [8, 25]. Many routine protocols used for the 2D analysis of cells, like cell count, metabolic activity, and viability can be transferred to 3D constructs. However, the protocols may necessitate higher reagent concentration or longer incubation times, due to diffusion limitations in 3D. Other procedures like flow cytometry, protein or RNA isolation may require the liberation of cells from the construct. Usually a combination of mechanical and enzymatic techniques is used for this purpose, although care must be taken not to damage cell receptors in these procedures, especially on prolonged treatment. Where enzymatic digestion is not available (i.e., in many synthetic polymers), researchers have introduced other degradation cues like light-activated degradation [116]. Sometimes, it is best to analyze the cells directly without extraction protocols, which always carry risk of altering or damaging the cells in some manner. Transparent materials are especially advantageous here, as they allow direct in situ imaging using (confocal) microscopy. If the material does not possess transparent optical properties, it can still be analyzed by employing methods traditionally used for the analysis of tissues like cryosectioning and immunostaining.

Existing materials for 3D cell culture keep evolving to meet new questions raised by researchers in the field. In most of the materials presented in this chapter, cell-instructive cues (like adhesions sequences) are either already present (e.g., in natural polymers like collagen) or engineered before the start of the experiment (e.g., RGD-adhesive sequences in PEG). While these "static" materials are quite valuable in their own right, they do not reflect the highly dynamic in vivo cell microenvironment and the role the ECM plays in cell signaling. Many researchers are, therefore, working on dynamic materials. These materials allow spatiotemporal control of a specific property like stiffness or availability of biological ligands. To achieve this, highly sophisticated material engineering is required, with reversible functionalization offering dynamic signal exposure to embedded cells [117]. Strategies used to achieve this include affinity-based swapping of bound biomolecules, enzyme-mediated systems, and photochemical reactions [118]. By the introduction of multiple orthogonal chemistries, it is even possible to provide independent control of several functionalities within the same material. In this way, it is possible to fine-tune material–cell interactions precisely, which is crucial for deepening our knowledge of the underlying mechanisms that govern physiological processes. Strong collaboration between material scientists, cell biologists, and engineers remains essential to achieve the aim of designing the next generation of materials for 3D cell culture, which will keep on striving to reflect the higher biological complexity of the ECM.

Take-Home Messages
- Metals, ceramics, composites, and polymers are all used for the culturing of cells, but polymers reign supreme as the most widely used materials in 3D cell culture, due to high design versatility.
- Natural biopolymers frequently have poor mechanical properties and exhibit batch-to-batch variations. Synthetic polymers are not cell instructive, as they do not contain inherent biological information for cells.
- Semi-synthetic or biohybrid polymers contain a synthetic part for mechanical control and a biological part for bioactivity.
- Biological matrices are derived from decellularized tissues. These provide the best approximation of ECM architecture and complexity. However, decellularized tissues are difficult to standardize and obtain.
- Cells can be encapsulated into the polymer network of a hydrogel, which requires a mild sol–gel reaction. There are many crosslinking mechanisms related to natural processes or engineered by chemists.
- Alternatively, the scaffold is first manufactured from polymers and then seeded with cells. Scaffolds are created to resemble the ECM by having a porous and/or fibrous structure.
- Important properties to consider in a 3D cell culture material are whether cells adhere to the material and whether mechanical properties are adequate for the purpose of experiment.

References

1. Frantz C, Stewart KM, Weaver VM. The extracellular matrix at a glance. J Cell Sci. 2010;123 (24):4195–200.
2. Liu H, Lin J, Roy K. Effect of 3D scaffold and dynamic culture condition on the global gene expression profile of mouse embryonic stem cells. Biomaterials. 2006;27(36):5978–89.
3. Hishikawa K, et al. Gene expression profile of human mesenchymal stem cells during osteogenesis in three-dimensional thermoreversible gelation polymer. Biochem Biophys Res Commun. 2004;317(4):1103–7.
4. Imamura Y, et al. Comparison of 2D-and 3D-culture models as drug-testing platforms in breast cancer. Oncol Rep. 2015;33(4):1837–43.
5. Longati P, et al. 3D pancreatic carcinoma spheroids induce a matrix-rich, chemoresistant phenotype offering a better model for drug testing. BMC Cancer. 2013;13(1):95.
6. Pickl M, Ries C. Comparison of 3D and 2D tumor models reveals enhanced HER2 activation in 3D associated with an increased response to trastuzumab. Oncogene. 2009;28(3):461–8.
7. Haycock JW. 3D cell culture: a review of current approaches and techniques. In: 3D cell culture. Totowa, NJ: Springer; 2011. p. 1–15.
8. Ruedinger F, et al. Hydrogels for 3D mammalian cell culture: a starting guide for laboratory practice. Appl Microbiol Biotechnol. 2015;99(2):623–36.

9. Seliktar D. Designing cell-compatible hydrogels for biomedical applications. Science. 2012;336 (6085):1124–8.
10. Stegemann JP, Hong H, Nerem RM. Mechanical, biochemical, and extracellular matrix effects on vascular smooth muscle cell phenotype. J Appl Physiol. 2005;98(6):2321–7.
11. Huh D, Hamilton GA, Ingber DE. From 3D cell culture to organs-on-chips. Trends Cell Biol. 2011;21(12):745–54.
12. Ravi M, et al. 3D cell culture systems: advantages and applications. J Cell Physiol. 2015;230 (1):16–26.
13. Carletti E, Motta A, Migliaresi C. Scaffolds for tissue engineering and 3D cell culture. In: 3D cell culture. Totowa, NJ: Springer; 2011. p. 17–39.
14. Lee J, Cuddihy MJ, Kotov NA. Three-dimensional cell culture matrices: state of the art. Tissue Eng Part B Rev. 2008;14(1):61–86.
15. Pham QP, Sharma U, Mikos AG. Electrospinning of polymeric nanofibers for tissue engineering applications: a review. Tissue Eng. 2006;12(5):1197–211.
16. Chiu Y-C, et al. Generation of porous poly (ethylene glycol) hydrogels by salt leaching. Tissue Eng Part C Methods. 2010;16(5):905–12.
17. Murphy SV, Atala A. 3D bioprinting of tissues and organs. Nat Biotechnol. 2014;32(8):773.
18. Lawrence BJ, Madihally SV. Cell colonization in degradable 3D porous matrices. Cell Adhes Migr. 2008;2(1):9–16.
19. Griffon DJ, et al. A comparative study of seeding techniques and three-dimensional matrices for mesenchymal cell attachment. J Tissue Eng Regen Med. 2011;5(3):169–79.
20. Tibbitt MW, Anseth KS. Hydrogels as extracellular matrix mimics for 3D cell culture. Biotechnol Bioeng. 2009;103(4):655–63.
21. Kirsch M, et al. Gelatin-Methacryloyl (GelMA) formulated with human platelet lysate supports mesenchymal stem cell proliferation and differentiation and enhances the hydrogel's mechanical properties. Bioengineering. 2019;6(3):76.
22. Pepelanova I, et al. Gelatin-Methacryloyl (GelMA) hydrogels with defined degree of functionalization as a versatile toolkit for 3D cell culture and extrusion bioprinting. Bioengineering. 2018;5(3):55.
23. Drury JL, Mooney DJ. Hydrogels for tissue engineering: scaffold design variables and applications. Biomaterials. 2003;24(24):4337–51.
24. Thiele J, et al. 25th anniversary article: designer hydrogels for cell cultures: a materials selection guide. Adv Mater. 2014;26(1):125–48.
25. Caliari SR, Burdick JA. A practical guide to hydrogels for cell culture. Nat Methods. 2016;13 (5):405.
26. Chevallay B, Herbage D. Collagen-based biomaterials as 3D scaffold for cell cultures: applications for tissue engineering and gene therapy. Med Biol Eng Comput. 2000;38(2):211–8.
27. Dasgupta A, Mondal JH, Das D. Peptide hydrogels. RSC Adv. 2013;3(24):9117–49.
28. Wang S, et al. Three-dimensional primary hepatocyte culture in synthetic self-assembling peptide hydrogel. Tissue Eng A. 2008;14(2):227–36.
29. Patterson J, Martino MM, Hubbell JA. Biomimetic materials in tissue engineering. Mater Today. 2010;13(1–2):14–22.
30. Andersen T, Auk-Emblem P, Dornish M. 3D cell culture in alginate hydrogels. Microarrays. 2015;4(2):133–61.
31. Abreu FO, et al. Influence of the composition and preparation method on the morphology and swelling behavior of alginate–chitosan hydrogels. Carbohydr Polym. 2008;74(2):283–9.
32. Weisel JW. The mechanical properties of fibrin for basic scientists and clinicians. Biophys Chem. 2004;112(2–3):267–76.

33. Schneider-Barthold C, et al. Hydrogels based on collagen and fibrin–frontiers and applications. BioNanoMaterials. 2016;17(1–2):3–12.
34. Willerth SM, et al. Optimization of fibrin scaffolds for differentiation of murine embryonic stem cells into neural lineage cells. Biomaterials. 2006;27(36):5990–6003.
35. Teixeira LSM, et al. Enzyme-catalyzed crosslinkable hydrogels: emerging strategies for tissue engineering. Biomaterials. 2012;33(5):1281–90.
36. Orban JM, et al. Crosslinking of collagen gels by transglutaminase. J Biomed Mater Res A. 2004;68(4):756–62.
37. Yung C, et al. Transglutaminase crosslinked gelatin as a tissue engineering scaffold. J Biomed Mater Res Part A. 2007;83(4):1039–46.
38. Hu B-H, Messersmith PB. Rational design of transglutaminase substrate peptides for rapid enzymatic formation of hydrogels. J Am Chem Soc. 2003;125(47):14298–9.
39. Khanmohammadi M, et al. Horseradish peroxidase-catalyzed hydrogelation for biomedical applications. Biomater Sci. 2018;6(6):1286–98.
40. Jin R, et al. Enzymatically crosslinked dextran-tyramine hydrogels as injectable scaffolds for cartilage tissue engineering. Tissue Eng A. 2010;16(8):2429–40.
41. Toh WS, et al. Modulation of mesenchymal stem cell chondrogenesis in a tunable hyaluronic acid hydrogel microenvironment. Biomaterials. 2012;33(15):3835–45.
42. Aguilar M, San Román J. Smart polymers and their applications. Cambridge: Elsevier; 2014.
43. Klouda L, Mikos AG. Thermoresponsive hydrogels in biomedical applications. Eur J Pharm Biopharm. 2008;68(1):34–45.
44. Bakaic E, et al. Injectable and degradable poly (oligoethylene glycol methacrylate) hydrogels with tunable charge densities as adhesive peptide-free cell scaffolds. ACS Biomater Sci Eng. 2017;4(11):3713–25.
45. Prabaharan M, Mano JF. Stimuli-responsive hydrogels based on polysaccharides incorporated with thermo-responsive polymers as novel biomaterials. Macromol Biosci. 2006;6 (12):991–1008.
46. Tatiana NM, et al. Hybrid collagen/pNIPAAM hydrogel nanocomposites for tissue engineering application. Colloid Polym Sci. 2018;296(9):1555–71.
47. Shim WS, et al. Biodegradability and biocompatibility of a pH-and thermo-sensitive hydrogel formed from a sulfonamide-modified poly (ε-caprolactone-co-lactide)–poly (ethylene glycol)–poly (ε caprolactone-co-lactide) block copolymer. Biomaterials. 2006;27(30):5178–85.
48. Jeong Y, et al. Enzymatically degradable temperature-sensitive polypeptide as a new in-situ gelling biomaterial. J Control Release. 2009;137(1):25–30.
49. Spicer CD, Pashuck ET, Stevens MM. Achieving controlled biomolecule–biomaterial conjugation. Chem Rev. 2018;118(16):7702–43.
50. Han J, Martinez BC, Ruan RR. Immobilization of Coleus blumei plant cells in temperature-sensitive hydrogel. Biotechnol Tech. 1996;10(5):359–62.
51. Wilems TS, et al. Effects of free radical initiators on polyethylene glycol dimethacrylate hydrogel properties and biocompatibility. J Biomed Mater Res A. 2017;105(11):3059–68.
52. Anjum F, et al. Tough, semisynthetic hydrogels for adipose derived stem cell delivery for chondral defect repair. Macromol Biosci. 2017;17(5):1600373.
53. Temenoff JS, et al. In vitro cytotoxicity of redox radical initiators for cross-linking of oligo (poly (ethylene glycol) fumarate) macromers. Biomacromolecules. 2003;4(6):1605–13.
54. Nuttelman CR, Tripodi MC, Anseth KS. In vitro osteogenic differentiation of human mesenchymal stem cells photoencapsulated in PEG hydrogels. J Biomed Mater Res A. 2004;68 (4):773–82.
55. Sharma B, et al. In vivo chondrogenesis of mesenchymal stem cells in a photopolymerized hydrogel. Plast Reconstr Surg. 2007;119(1):112–20.

56. Xiong Z, et al. Femtosecond laser induced densification within cell-laden hydrogels results in cellular alignment. Biofabrication. 2019;11(3):035005.
57. Tsang VL, et al. Fabrication of 3D hepatic tissues by additive photopatterning of cellular hydrogels. FASEB J. 2007;21(3):790–801.
58. Zhu W, et al. 3D bioprinting mesenchymal stem cell-laden construct with core–shell nanospheres for cartilage tissue engineering. Nanotechnology. 2018;29(18):185101.
59. Benoit DS, Anseth KS. Heparin functionalized PEG gels that modulate protein adsorption for hMSC adhesion and differentiation. Acta Biomater. 2005;1(4):461–70.
60. Shih H, Lin C-C. Photoclick hydrogels prepared from functionalized cyclodextrin and poly (ethylene glycol) for drug delivery and in situ cell encapsulation. Biomacromolecules. 2015;16 (7):1915–23.
61. Gong H, et al. 3D-engineered GelMA conduit filled with ECM promotes regeneration of peripheral nerve. J Biomed Mater Res A. 2020;108(3):805–13.
62. Gao Y, Jin X. Dual crosslinked methacrylated alginate hydrogel micron fibers and tissue constructs for cell biology. Mar Drugs. 2019;17(10):557.
63. Antunes J, et al. In-air production of 3D co-culture tumor spheroid hydrogels for expedited drug screening. Acta Biomater. 2019;94:392–409.
64. Hu X, Li D, Gao C. Chemically cross-linked chitosan hydrogel loaded with gelatin for chondrocyte encapsulation. Biotechnol J. 2011;6(11):1388–96.
65. Gevaert E, et al. Galactose-F unctionalized gelatin hydrogels improve the functionality of encapsulated Hepg2 cells. Macromol Biosci. 2014;14(3):419–27.
66. Poon YF, et al. Hydrogels based on dual curable chitosan-graft-polyethylene glycol-graft-methacrylate: application to layer-by-layer cell encapsulation. ACS Appl Mater Interfaces. 2010;2(7):2012–25.
67. Li L, et al. Biodegradable and injectable in situ cross-linking chitosan-hyaluronic acid based hydrogels for postoperative adhesion prevention. Biomaterials. 2014;35(12):3903–17.
68. Prince E, et al. Patterning of structurally anisotropic composite hydrogel sheets. Biomacromolecules. 2018;19(4):1276–84.
69. Gendler E, Gendler S, Nimni M. Toxic reactions evoked by glutaraldehyde-fixed pericardium and cardiac valve tissue bioprosthesis. J Biomed Mater Res. 1984;18(7):727–36.
70. Korzhikov VA, et al. Water-soluble aldehyde-bearing polymers of 2-deoxy-2-methacrylamido-d-glucose for bone tissue engineering. J Appl Polym Sci. 2008;108(4):2386–97.
71. Xu K, et al. Thiol-ene Michael-type formation of gelatin/poly (ethylene glycol) biomatrices for three-dimensional mesenchymal stromal/stem cell administration to cutaneous wounds. Acta Biomater. 2013;9(11):8802–14.
72. Liu ZQ, et al. Dextran-based hydrogel formed by thiol-Michael addition reaction for 3D cell encapsulation. Colloids Surf B: Biointerfaces. 2015;128:140–8.
73. Yu Y, et al. Thiolated gellan gum hydrogels as a peptide delivery system for 3D neural stem cell culture. Mater Lett. 2020;259:126891.
74. Liu Z, et al. Spatiotemporally controllable and cytocompatible approach builds 3D cell culture matrix by photo-uncaged-thiol Michael addition reaction. Adv Mater. 2014;26(23):3912–7.
75. Mūnoz Z, Shih H, Lin C-C. Gelatin hydrogels formed by orthogonal thiol–norbornene photo-chemistry for cell encapsulation. Biomater Sci. 2014;2(8):1063–72.
76. Bian S, et al. The self-crosslinking smart hyaluronic acid hydrogels as injectable three-dimensional scaffolds for cells culture. Colloids Surf B: Biointerfaces. 2016;140:392–402.
77. Madl CM, Heilshorn SC. Bioorthogonal strategies for engineering extracellular matrices. Adv Funct Mater. 2018;28(11):1706046.

78. Agard NJ, Prescher JA, Bertozzi CR. A strain-promoted [3+ 2] azide– alkyne cycloaddition for covalent modification of biomolecules in living systems. J Am Chem Soc. 2004;126 (46):15046–7.

79. Fu S, et al. Injectable hyaluronic acid/poly (ethylene glycol) hydrogels crosslinked via strain-promoted azide-alkyne cycloaddition click reaction. Carbohydr Polym. 2017;169:332–40.

80. Truong VX, et al. In situ-forming robust chitosan-poly (ethylene glycol) hydrogels prepared by copper-free azide–alkyne click reaction for tissue engineering. Biomater Sci. 2014;2(2):167–75.

81. Piluso S, et al. Sequential alkyne-azide cycloadditions for functionalized gelatin hydrogel formation. Eur Polym J. 2018;100:77–85.

82. Owen SC, et al. Hyaluronic acid click hydrogels emulate the extracellular matrix. Langmuir. 2013;29(24):7393–400.

83. Nahm D, et al. A versatile biomaterial ink platform for the melt electrowriting of chemically-crosslinked hydrogels. Mater Horiz. 2020;7(3):928–33.

84. Gilbert TW, Sellaro TL, Badylak SF. Decellularization of tissues and organs. Biomaterials. 2006;27(19):3675–83.

85. Hoshiba T, et al. Decellularized matrices for tissue engineering. Expert Opin Biol Ther. 2010;10 (12):1717–28.

86. Song JJ, Ott HC. Organ engineering based on decellularized matrix scaffolds. Trends Mol Med. 2011;17(8):424–32.

87. Carter P, Bhattarai N. Chapter 7. Bioscaffolds: fabrication and performance. In: Engineered biomimicry. Oxford: Elsevier; 2013.

88. Heijkants R, et al. Polyurethane scaffold formation via a combination of salt leaching and thermally induced phase separation. J Biomed Mater Res Part A. 2008;87(4):921–32.

89. Nam YS, Park TG. Porous biodegradable polymeric scaffolds prepared by thermally induced phase separation. J Biomed Mater Res. 1999;47(1):8–17.

90. Li S, Wang K, Li M. Morphology and pore size distribution of biocompatible interconnected porous poly (L-lactic acid) foams with nanofibrous structure prepared by thermally induced liquid–liquid phase separation. J Macromol Sci Part B. 2010;49(5):897–919.

91. Averianov I, Korzhikov V, Tennikova T. Synthesis of poly (lactic acid) and the formation of poly (lactic acid)-based supraporous biofunctional materials for tissue engineering. Polym Sci Ser B. 2015;57(4):336–48.

92. Zare-Mehrjardi N, et al. Differentiation of embryonic stem cells into neural cells on 3D poly (D, L-lactic acid) scaffolds versus 2D cultures. Int J Artif Organs. 2011;34(10):1012–23.

93. Ma PX, Zhang R. Synthetic nano-scale fibrous extracellular matrix. J Biomed Mater Res. 1999;46(1):60–72.

94. Holzwarth JM, Ma PX. Biomimetic nanofibrous scaffolds for bone tissue engineering. Biomaterials. 2011;32(36):9622–9.

95. Kirsebom H, et al. Macroporous scaffolds based on chitosan and bioactive molecules. J Bioact Compat Polym. 2007;22(6):621–36.

96. Zhao Y, Chen Z, Wu T. Cryogelation of alginate improved the freeze-thaw stability of oil-in-water emulsions. Carbohydr Polym. 2018;198:26–33.

97. Liu X, Ma PX. Phase separation, pore structure, and properties of nanofibrous gelatin scaffolds. Biomaterials. 2009;30(25):4094–103.

98. Kao H-H, et al. Preparation of gelatin and gelatin/hyaluronic acid cryogel scaffolds for the 3D culture of mesothelial cells and mesothelium tissue regeneration. Int J Mol Sci. 2019;20 (18):4527.

99. Lozinsky VI. Cryogels on the basis of natural and synthetic polymers: preparation, properties and application. Russ Chem Rev. 2002;71(6):489–511.

100. Park J, et al. Clinical application of bone morphogenetic protein-2 microcarriers fabricated by the cryopolymerization of gelatin methacrylate for the treatment of radial fracture in two dogs. In Vivo. 2018;32(3):575–81.
101. Repanas A, et al. Coaxial electrospinning as a process to engineer biodegradable polymeric scaffolds as drug delivery systems for anti-inflammatory and anti-thrombotic pharmaceutical agents. Clin Exp Pharmacol. 2015;5:192.
102. Gryshkov O, et al. Advances in the application of electrohydrodynamic fabrication for tissue engineering. J Phys Conf Ser. 2019;1236:012024.
103. Chu PK, Liu X. Biomaterials fabrication and processing handbook. Boca Raton, FL: CRC Press; 2008.
104. Wang X, Ding B, Li B. Biomimetic electrospun nanofibrous structures for tissue engineering. Mater Today. 2013;16(6):229–41.
105. Damanik FF, et al. Biological activity of human mesenchymal stromal cells on polymeric electrospun scaffolds. Biomater Sci. 2019;7(3):1088–100.
106. Hajzamani D, et al. Effect of engineered PLGA-gelatin-chitosan/PLGA-gelatin/PLGA-gelatin-graphene three-layer scaffold on adhesion/proliferation of HUVECs. Polym Adv Technol. 2020;31(9):1896–910.
107. Unal S, et al. Glioblastoma cell adhesion properties through bacterial cellulose nanocrystals in polycaprolactone/gelatin electrospun nanofibers. Carbohydr Polym. 2020;233:115820.
108. Zhang Y, et al. Characterization of the surface biocompatibility of the electrospun PCL-collagen nanofibers using fibroblasts. Biomacromolecules. 2005;6(5):2583–9.
109. Kook Y-M, et al. Bi-compartmental 3D scaffolds for the co-culture of intervertebral disk cells and mesenchymal stem cells. J Ind Eng Chem. 2016;38:113–22.
110. Majidi SS, et al. Wet electrospun alginate/gelatin hydrogel nanofibers for 3D cell culture. Int J Biol Macromol. 2018;118:1648–54.
111. Feng Z-Q, et al. The effect of nanofibrous galactosylated chitosan scaffolds on the formation of rat primary hepatocyte aggregates and the maintenance of liver function. Biomaterials. 2009;30 (14):2753–63.
112. Khanmohammadi M, et al. Cell encapsulation in core-shell microcapsules through coaxial electrospinning system and horseradish peroxidase-catalyzed crosslinking. Biomed Phys Eng Express. 2020;6(1):015022.
113. Guo Y, et al. Modified cell-electrospinning for 3D myogenesis of C2C12s in aligned fibrin microfiber bundles. Biochem Biophys Res Commun. 2019;516(2):558–64.
114. Yeo M, Kim G. Fabrication of cell-laden electrospun hybrid scaffolds of alginate-based bioink and PCL microstructures for tissue regeneration. Chem Eng J. 2015;275:27–35.
115. Yeo M, Kim GH. Anisotropically aligned cell-laden nanofibrous bundle fabricated via cell electrospinning to regenerate skeletal muscle tissue. Small. 2018;14(48):1803491.
116. Kloxin AM, et al. Photodegradable hydrogels for dynamic tuning of physical and chemical properties. Science. 2009;324(5923):59–63.
117. DeForest CA, Anseth KS. Advances in bioactive hydrogels to probe and direct cell fate. Annu Rev Chem Biomol Eng. 2012;3:421–44.
118. Hammer JA, West JL. Dynamic ligand presentation in biomaterials. Bioconjug Chem. 2018;29 (7):2140–9.

Hydrogels for 3D Cell Culture

5

Antonina Lavrentieva and Jane Spencer-Fry

Contents

A. Lavrentieva (✉)
Institute of Technical Chemistry, Leibniz University of Hannover, Hannover, Germany
e-mail: Lavrentieva@iftc.uni-hannover.de

J. Spencer-Fry
Apollo Technology Partners Ltd., Pangbourne, UK
e-mail: jsf@apollotp.com

© Springer Nature Switzerland AG 2021
C. Kasper et al. (eds.), *Basic Concepts on 3D Cell Culture*, Learning Materials in Biosciences,
https://doi.org/10.1007/978-3-030-66749-8_5

What You Will Learn in This Chapter
One of the fast growing 3D mammalian cell culture platforms are hydrogels—three-dimensional, crosslinked networks of polymers. In this chapter we describe the fundamentals of hydrogels and provide an overview of sources from which hydrogels can be derived, such as animal, non-animal, synthetic, or combinations. The physical/mechanical requirements of hydrogels are discussed in order that they produce a physiologically relevant environment for 3D cell cultivation. We review the characterization methods used for hydrogels and how this impacts application in 3D cell culture and regenerative medicine. Modification of hydrogels by crosslinking affords them the property of tunability and degradation to recover cells for downstream analysis provides an opportunity to gather additional supportive data, these areas are examined along with the physical properties needed for optimum use in cell monitoring and analysis techniques. Creation of gradient hydrogels and incorporation into microfluidic and organ-on-a-chip models opens the possibility to recapitulate more closely the in vivo environment. We cover the areas in which hydrogels are being applied to support 3D cell culture such as bioprinting and bioprocessing and discuss the future potential of these versatile materials in taking cell research from the bench to the clinic.

5.1   Introduction: What Is a Hydrogel?

Hydrogels are three-dimensional network structures permeable to oxygen and nutrients and capable of taking on large amounts of water, which make them attractive for use in biological applications. Presence of chemical or physical crosslinks and/or chain entanglements typically prevents the hydrogel from dissolving, therefore retaining structure and stiffness. They can be manufactured synthetically or extracted from natural sources, e.g. collagen, gelatin, alginate, and nanofibrillar cellulose. When used for 3D cell culture the hydrogel properties can be adapted to match the specific use, important as different body tissues have different physical and biochemical requirements. The inherent and versatile properties of hydrogels have seen them used in many applications including controlled drug delivery systems, biosensors, tissue engineering scaffolds, artificial organs, wound healing bandages, physiological membranes, contact lenses, and microfluidic valves.

In cell culture hydrogels were initially used to coat tissue culture vessels providing a 2.5D environment for adherent cell growth. With a drive to bridge the gap between in vitro and in vivo conditions there has been a shift away from the 2D model toward 3D to create more human relevant data. Hydrogels have been the natural choice for development of these new 3D cell culture systems.

5.2   Hydrogel Classification

There are different categories into which hydrogels can be classified: (1) *structural composition* (e.g., homopolymers, copolymers, or interpenetrating networks), (2) *origin/source of polymers* (e.g., natural, semi-synthetic, or synthetic hydrogels), (3) *crosslinking* (e.g., photo, physically or chemically crosslinked hydrogels), (4) *responsiveness* to stimuli (e.g., temperature-responsiveness and pH-responsiveness), (5) *molecular charge* (cationic, anionic, neutral, ampholytic), and (6) *crosslinking reversibility* (reversibly or irreversibly crosslinked hydrogels) [1–3] (Fig. 5.1). Additionally, hydrogels can be classified according to their design features (physical, biological, or mass-transport design features) [3].

5.2.1   Structural Composition

Classification by structural composition divides hydrogels into *homopolymeric* (derived from the single monomer species), copolymeric (derived from two or more monomer species), and multipolymeric (also called interpenetrating networks), derived from two independent crosslinked polymers, where at least one of them being synthesized and/or crosslinked within the immediate presence of the other and without any covalent bonds between them [4].

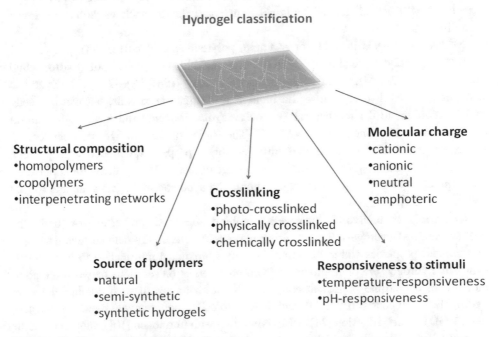

Fig. 5.1 Variants of hydrogel classifications

The most widely used homopolymeric hydrogels in 3D mammalian cell culture are collagen [5], fibrin [6], and nanofibrillar cellulose [7]. Gelatin methacryloyl (GelMA) [8, 9] and PEG-fibrinogen [10, 11] hydrogels are macromonomeric homopolymer hydrogels. Copolymeric hydrogels suitable for 3D cell culture are represented by alginates [12], hyaluronic acid and poly(N-isopropolyacrylamide) hydrogels [13], PEG-based copolymers (PEGMEMA–MEO₂MA–PEGDA) [14], and synthetic saccharide–peptide hydrogels [15]. Some examples of multipolymeric hydrogels are networks of dextran and gelatin [16], gelatin and silk fibroin [17], or alginate and reconstituted basement membrane matrix hydrogels [18].

5.2.2 Origin of Polymers

Natural hydrogels are made of polysaccharides (alginate, agarose, glucan, hyaluronic acid, nanocellulose, and chitosan) or proteins (collagen, albumin, fibrin, and silk proteins), derived by extractions from biological sources. Collagen, fibrin, and hyaluronic acid are natural constituents of the ECM, while alginate and agarose are derived from marine algae. Another new natural source of hydrogels is nanofibrillar cellulose, which is extracted from wood [19]. The major advantage of natural hydrogels is their biocompatibility and closest proximity to the in vivo cell microenvironment. Animal ECM-derived hydrogels perfectly support cell adhesion, while hydrogels of non-animal (non-human) origin are readily available and avoid possible viral contamination. There are a few notable disadvantages associated with animal derived hydrogels, such as batch-to-batch variation and lower tunability [20].

Synthetic hydrogels include, for example, poly(ethylene glycol) (PEG), poly(acrylic acid) (PAA), poly(ethylene oxide) (PEO), poly(vinyl alcohol) (PVA), poly(hydroxyethyl methacrylate) (PHEMA), and poly(methacrylic acid) (PMMA) [20, 21]. Synthetic structures of such hydrogels offer no biological information to cells, but can be easily tuned according to the mechanical (viscoelastic) requirements and have high uniform quality as well as defined structure [3, 22]. The choice of the hydrogel is dependent on the experimental setup (i.e., required stiffness, optical properties, conductive properties), material availability, and cost. There is no shortage of materials to evaluate for application development, given the great variety of natural and synthetic hydrogels available for 3D cell culture [23].

Semi-synthetic biohybrid hydrogels combine the best properties of both the abovementioned hydrogels. The resulting hydrogels have highly defined technical (and mechanical) properties and at the same time offer biological information to cells. Semi-synthetic hydrogels are often produced in co-polymerization reactions between the polymer precursor and the biological conjugate. Some examples of such combinations are hydrogels which consist of PEG (which is inert to cells) modified with peptide sequences (e.g., RGD) for cell adhesion [24] or PEG modified with fibrinogen [10]. Another example of biohybrid hydrogels is GelMA—here, cell promoting gelatin is derivatized with methacrylamide and methacrylate groups, which provide the hydrogel with shape fidelity and stability at physiological temperature [8].

5.2.3 Crosslinking

The crosslinks between individual polymer molecules maintain the entire 3D structure of the hydrogel after swelling in water. For use in 3D cell culture (especially in the case of cell encapsulation prior polymerization) not only the polymer material, but also crosslinking reaction must be cell-friendly, which means that reaction conditions, substrates, and products should not affect cell viability. One of the widely used strategies is crosslinking with visible or UV light through *photopolymerization* of acrylate or methacrylate-modified hydrogel polymers. Here, photo-initiator is used to create free radicals which attack the vinyl groups of precursor molecules, resulting in covalent crosslinking of the hydrogel within seconds or minutes upon irradiation [8, 25]. Some hydrogels can be crosslinked by *physical methods*, such as ionic crosslinking of alginate [26], thermally induced gelation [27], or self-assembling amphiphiles [28]. Hydrogels can be *covalently crosslinked* in polymerization reactions (see also Chap. 5 "Biological, natural and synthetic 3D matrices"), which involve gentle chemistries under physiological conditions (bio-orthogonal chemistry). Hydrogels can be also crosslinked *enzymatically*. Here, transglutaminase is widely used to crosslink peptide-functionalized hydrogel materials [29, 30].

Choosing the polymerization strategy of the hydrogel for use in 3D cell culture, the researcher must also take into account the time of polymerization (gelation kinetics)—some polymerization reactions are too fast to ensure an even cell distribution and some are too slow, so that cells sediment to the bottom of the construct before complete polymerization.

5.2.4 Stimuli-Responsive Hydrogels

Although 3D structure of hydrogels brings in vitro cell culture closer to the physiological in vivo conditions, static materials cannot fully mimic the dynamicity of native microenvironment [31]. Stimuli-responsive hydrogels can change their physical and chemical properties depending on the external stimuli, which makes them an important tool for basic research and biomedical applications. Depending on the ability to react to these stimuli, hydrogels can be divided into *pH*-responsive [32], *temperature*-responsive [33], *light/photo*-responsive [34], and *electric field*-responsive hydrogels [35]. These hydrogels can provide cells with irreversible or reversible spatiotemporal modulation of cues, directing cell behavior [31].

5.2.5   Molecular Charge and Reversibility of Crosslinking

Finally, hydrogels can be sorted by molecular charge and reversibility of crosslinking. Bilayer phospholipid membranes of cells are negatively charged and positively charged *cationic hydrogels* can facilitate cell attachment [36]. *Anionic hydrogels* have been shown to induce formation of the bone mineral hydroxyapatite by the cells [37] and can be used as a bone regeneration matrix [38].

Reversibility of crosslinking plays an important role in cell recovery and analysis. If hydrogels are crosslinked chemically the junction points are usually permanent covalent bonds [39]. If such hydrogels cannot be degraded by the cells, it limits cell spreading and migration. Hydrogels crosslinked physically are usually reversible. So, alginates, for example, can be easily dissociated by calcium chelators (e.g., EDTA and sodium citrate).

5.3   Physical Requirements for Cell Culture

The extracellular matrix (ECM) surrounds most cells in tissues of complex organisms, protecting them from stress and regulating cellular functions such as spreading, migration, proliferation, and stem cell differentiation. Stiffness of the ECM is considered to have implications for development, differentiation, disease, and regeneration [40].

In Fig. 5.2 the relationship between ECM stiffness and cell type is depicted. There is a large variation in the in vivo ECM environment with neural cells at the softer end and cartilage and bone cells at the stiffer end of the range. Studies have shown that by adjusting the stiffness of the matrix rather than making changes just to the biochemical environment (i.e., use of growth factors or defined media) then directed differentiation can be achieved. Matching the stiffness of the hydrogel to the tissue is of interest particularly when targeting MSC fates, since MSCs (and numerous other cell types) can convert external mechanical clues to intracellular biochemical signals. This ability to sense mechanical microenvironment called *mechanosensing* is described in several studies and reviews [41, 42]. Those MSCs cultured in lower stiffness hydrogel (2 kPa) show a tendency to differentiate toward cells expressing neural markers; those cultured in hydrogel with a kPa of 10 formed myocytes and those cultured on rigid substrates (40 kPa) become osteoblasts [43].

Fig. 5.2 Illustration of ECM stiffness (elastic modulus) versus cell type

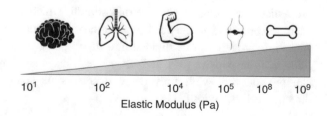

10^1 10^2 10^4 10^5 10^8 10^9

Elastic Modulus (Pa)

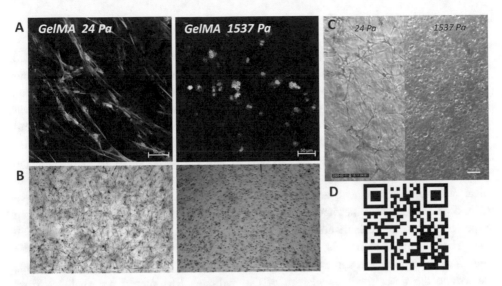

Fig. 5.3 Microscopic analysis of hAD-MSCs and HUVECs co-culture after 3 days of cultivation in GelMA hydrogels with stiffness of 24 Pa and 1537 Pa: (**a**) Confocal microscopy pictures (green—hAD-MSCs, red—HUVECs), (**b**) microphotographs, (**c**) screenshot of time-lapse, and (**d**) QR code for time-lapse video of hAD-MSCs spreading

There are many examples of studies where matrix stiffness has been shown to play a role in cell development, migration or differentiation, for example, neural cells, MSC differentiation, muscle cells, breast cancer cells, and bone [43–47].

Anchorage-dependent cells are highly responsive to hydrogel properties (stiffness and pore structure) and encapsulated cells demonstrate higher spreading in low stiffness hydrogels and no spreading by high stiffness [8, 48, 49]. So, MSCs show good spreading already starting on day 1 after encapsulation in Gelatin-Methacryloyl (GelMA) hydrogel with a low degree of functionalization (final stiffness 24 Pa) and no spreading in the same hydrogel with stiffness of 1537 Pa (Fig. 5.3 and supplementary Video 1).

It should be noted that if the pore size within the hydrogel is too small or the hydrogel cannot be proteolytically degraded by the cells, anchorage-dependent cells will not survive long. In addition, stability of the hydrogel is of importance as the matrix needs to be able to withstand standard cell culture operations, such as transfer to and from the microscope and media change without loss or degradation for the duration of the experiment. Ideally the best case would be a slowly biodegradable hydrogel which can be replaced by de novo formed ECM.

5.4   Material Characterization

Precise control of hydrogel properties belongs to the essential routines of hydrogel-based 3D cell culture. As already mentioned, most cells are sensitive to the mechanical microenvironment and knowledge as well as control of the mechanical properties of a hydrogel, like stiffness or viscoelasticity, plays a crucial role in the establishment of desired cultivation conditions. Hydrogel mechanical properties and polymerization dynamics (gelation) can be characterized using *rheology* [3]. Using only relatively small sample volumes (100–1000 µL), modern rheometers can quickly and sensitively measure the mechanical properties of hydrogels. The hydrogel is placed between parallel plates (alternatively cone-plate or concentric cylinders) and torsional oscillation generates shear flow in the sample (Fig. 5.4). Protocols for the rheological characterization of hydrogels and different sweep experiments are well-established and described [50]. *Time sweep experiments* determine the gelation time of hydrogels, *strain sweep experiments* measure the linear viscoelastic region of the hydrogel in dependency to the applied strain. *Frequency sweep experiments* determine the linear modulus plateau of the hydrogel. Rheometers are available from various manufacturers (Anton Paar, TA Instruments, Malvern or Thermo Fisher). Typical equipment for hydrogel characterization is a rotational- and oscillatory rheometer with

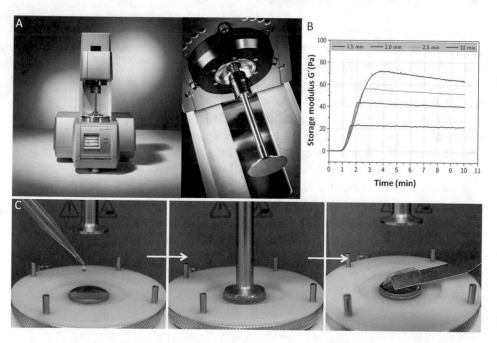

Fig. 5.4 Rheological characterization of hydrogels: (**a**) rheometer, (**b**) a time-sweep experiment of a photo-crosslinkable hydrogel, showing storage modulus curves resulting from different illumination times, and (**c**) measurement of the polymerization of photosensitive hydrogels in a parallel plate system

parallel plate geometry, Peltier-element (for precise temperature settings), and a UV-curing system (for UV photo-crosslinkable hydrogels). Precise temperature settings are crucial for characterization of hydrogels with crosslinking via temperature transition or enzymatic crosslinking [51]. Temperature transition, enzymatic or photo-crosslinking reactions trigger the hydrogel development from its original liquid state to its fully polymerized state (sol-gel transition). Major viscoelastic properties of hydrogels are the *storage modulus* (G'), which measures the stiffness, and the *loss modulus* (G"), reflecting the hydrogel viscosity [52]. Taken together, G' and G" represent the shear modulus G of a hydrogel, according to Eq. (5.1):

$$G = \sqrt{(G')^2 + (G'')^2} \qquad (5.1)$$

Another important characteristic of hydrogels is their *swelling* behavior. Hydrogel swelling characteristics influence the materials' mechanical properties, shape fidelity and diffusion of nutrients, and depend on crosslinking density, hydrophilicity of the polymer, and interactions with medium or other solvent [53]. For determination of swelling, the polymerized sample is placed into solvent/medium for 24 h until equilibrium, weighed, freeze-dried, and re-weighed again. The mass-swelling ratio is calculated as the ratio of swollen hydrogel mass to the mass of dry material [54]. The swelling degree of hydrogels is usually inversely proportional to the hydrogel concentration and degree of crosslinking—the higher the crosslinking, the lower the swelling.

Structural characteristics of hydrogels can be evaluated by several techniques: *scanning electron microscopy (SEM)*, *cryosectioning*, or *confocal microscopy* of fluorescently stained hydrogels. SEM micrographs of the three-dimensional polymer network and the pores of hydrogels provide information about morphological structure and pore architecture—here the effect of modifications can be estimated on the hydrogel pores [55]. For SEM, hydrogels are usually first swollen, then frozen in liquid nitrogen, freeze-dried, and sputtered with gold prior to the observation. For cryosectioning, hydrogels are frozen in the optimal cutting temperature compound (Tissue-Tek®) and sections are prepared and collected on slides using cryostat [56]. Additionally, *atomic force microscopy (AFM)* is used for high-resolution characterization of hydrogel topography, as well as for probing the elastic modulus (elastic moduli map), disclosing the surface roughness and stiffness of the hydrogel constructs [57].

5.5 Gradient Hydrogels

Oxford dictionary defines gradients as "an increase or decrease in the magnitude of a property (e.g., temperature, pressure, or concentration) observed in passing from one point or moment to another" (https://www.lexico.com/definition/gradient). In vivo, gradients are the essential part of all living organisms, beginning already in the early embryogenesis as

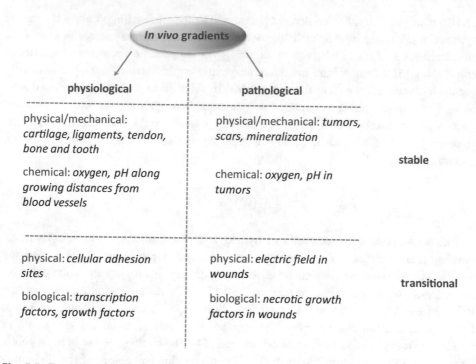

Fig. 5.5 Examples of in vivo gradients

the gradient distribution of the transcriptional factors. In all multicellular organisms (or colonies of unicellular organisms) gradients of different nature and temporal resolution can be found. Gradients can be *stable* or *transitional*, *physiological* or *pathological*. By the nature, gradients can be classified as *physical*, *chemical*, and *biological* (Fig. 5.5). Stable physiological mechanical gradients can be found in different tissues like cartilage, ligaments, tendon, bone, and tooth [58–60]. Stable chemical physiological gradients (oxygen concentrations, pH gradient as a result of catabolite distribution) go along the increasing distances from blood vessels [61]. Stable pathological gradients (chemical, physical, and biological) can be found in tumors, where fast growing cellular mass breaks tissue homeostasis [62, 63]. Transitional physiological gradients direct embryonic development, growth of blood vessels, and tissues [64–66]. Transitional pathological gradients are present in wounds, scars, during mineralization of the artery walls or fibrogenesis in kidney [59, 67, 68].

Cellular fate is strongly influenced by the composition of the tissue microenvironment and most cell types sense physical, chemical, and biological characteristics of their external microenvironment and convert them to intracellular biochemical signals. The influence of different microenvironmental signals, alone or in combination with each other, can be studied in hydrogel-based 3D cultivation systems. The above discussed tunable properties of hydrogels allow not only creation of desired in situ mechanical, biological, and architectural microenvironments, but also give the opportunities to create gradients inside of the bulk hydrogel constructs. Fabrication of gradients in hydrogels (1) enables *the recapitulation of in vivo gradients* and (2) can help to *find an optimal niche* for different cell types and co-cultures [69]. Similarly to the in vivo gradients, gradient hydrogels can be divided into three major groups: *physical* (mechanical properties of material), *biological* (bioactive molecules incorporation), and *chemical* (material composition) gradient hydrogels (Fig. 5.6) [69, 70]. Gradients in hydrogels can be continuous or stepped. By profiles gradients are divided into linear, radial, exponential, or sigmoidal [70].

There are many different methods to create gradient hydrogels, some of them are presented in Fig. 5.7. Mechanical (hydrogel stiffness and pore architecture) gradients can be created by two main strategies: (1) variation of crosslinker concentration in the pre-polymer solution and (2) variation of polymerization intensity. Variation of crosslinker concentration can be made by dynamic mixing with the help of *two-syringe pump system*

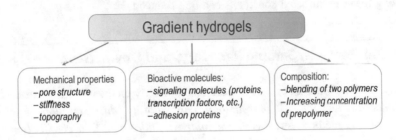

Fig. 5.6 Types of gradient hydrogels

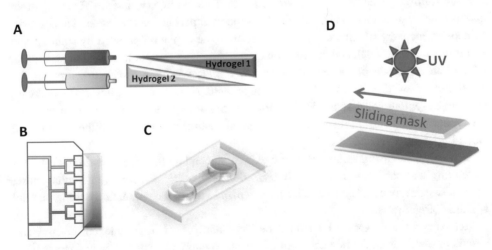

Fig. 5.7 Selected examples of gradient hydrogel production: (**a**) dynamic mixing of two precursors with, e.g., two-syringe system, (**b**) microfluidic tree mixer, (**c**) limited mixing in the Hele-Shaw cell device, and (**d**) sliding mask technique

[49, 71, 72], *microfluidic* techniques [73–76], or limited mixing in the *Hele-Shaw* cell device [77] (Fig. 5.7). Variation of polymerization can be created (in the case of photopolymerized hydrogels) by the use of *sliding mask* [78] or photolithographic patterning [79].

Most of the abovementioned techniques can be also used to create biological and chemical gradients [69]. So, e.g., differentiation factors such as bone morphogenic protein 2 (BMP-2) and transforming growth factor ß1 (TGF-ß1), incorporated in heparin-alginate hydrogels in opposite directions (Fig. 5.3a) led to higher osteogenic differentiation of mesenchymal stem cells along increasing BPM-2 and higher chondrogenic differentiation in the direction of the TGF-ß1 concentration growth [72]. Such biological gradient hydrogels were also engineered for in vitro disease model application, like gradients of epidermal growth factor to study of tumor cell intravasation [80]. Moreover, new methods to fabricate hydrogels with *combined multiple gradients* of different natures were reported [81]. Using these combinatory gradients, complex disease models can be created and better functioning tissue engineered constructs can be produced.

5.6   Cell Analysis, Sample Recovery, and Downstream Analysis

Different analytical techniques may be performed directly on the in vitro 3D disease model or TE construct such as cell *morphology, cell viability, differentiation* or expression biomarkers using methods such as *phase contrast, fluorescence,* or *confocal microscopy* [82]. However, this non-invasive monitoring often does not provide the complete picture and supplemental data obtained from RNA isolation, protein extraction, single cell isolation and following analysis by, e.g., Western blot, flow cytometry, and qPCR among other downstream applications may be required to support experimental findings and hypothesis. Moreover, recovery of spheroids, organoids, or tissue explants followed by staining and sectioning can reveal detailed information on cell structure, morphology, and organization, which can be invaluable when attempting to replicate in vivo tumor structure for the development of disease models, new drugs and/or treatment regimes. Downstream techniques require cells, spheroid, organoid, or tissue explants to be recovered quickly, easily, and intact with no residual matrix present that may interfere with the process and data ultimately generated.

The sample recovery technique employed will depend on the type of hydrogel used for the culture and can range from *depolymerization, enzymatic digestion* (e.g., collagenase, trypsin, dispase, or cellulase), and *mechanical processes,* or a combination used in parallel which is often the case.

Recovery of cells grown on animal derived matrices such as collagen, gelatin, and basement matrix is most often achieved by the use of enzymatic digestion and mechanical agitation. For example, recovery from Matrigel® can be performed by use of proteases that depolymerize the matrix within a few hours on ice using gentle agitation via a flatbed shaker. The sample is then washed several times with PBS and cells pelleted. Alternatively, dispase, a metalloenzyme which gently releases cells, can be used in combination with mechanical agitation. Protocols for recovery of cells from collagen recommend use of a collagenase/dispase solution where the sample is pipetted up and down to break up the gel completely, followed by addition of an EDTA/EGTA-containing solution to quench the reaction. Care should be taken to ensure the correct type of collagenase is used as this may impact cell viability should further culture of recovered cells be required.

Hyaluronic acid (HA) hydrogel from thiol-modified HA can be returned to solution phase by addition of dithiothreitol as demonstrated with L-929 murine fibroblasts [83].

Protocols for PEG-based hydrogels employ an enzyme α-chymotrypsin to release spheroids from the matrix in combination with mechanical shaking [84]. Block copolymers based on disulfide-containing polyethylene glycol diacrylate crosslinkers have been shown to be dissociated using the thiol–disulfide exchange reaction in the presence of N-acetyl-cysteine or glutathione, this dissolves the hydrogel network and cells recovered by centrifugation [85]. Examples of cells recovered in this manner include murine NIH 3T3 fibroblasts, human HepG2 C3A hepatocytes, human bone marrow-derived mesenchymal stem cells (MSCs), and human umbilical vein endothelial cells (HUVECs).

When recovering cells from a natural non-animal product such as nanocellulose, cellulase enzymes may be used to digest the cellulose fibers [86]. These enzymes break the cellulose fibers into glucose molecules removing the hydrogel structure to form a solution. The digestion can be done in situ without mechanical agitation, which is an advantage when trying to preserve cell structures for sectioning. Ionic alginate hydrogels require the addition of chelating agents (e.g., EDTA and sodium citrate) to reverse the crosslinking and release the encapsulated cells [87].

5.7   Technologies into Which Hydrogels Can be Incorporated

All of the hydrogel properties described above make them useful tools in a wide variety of applications. In 3D cell culture and regenerative medicine hydrogels are widely used as bioinks in *bioprinting* (Chap. 11). Here, cells resuspended in unpolymerized hydrogels are printed in 3D structures, which allow precise control of the 3D construct geometry and spatial cell distribution. Hydrogel polymerization, if required for stabilization of the construct, takes place during or directly after printing. Another frequently used application of hydrogels is their implementation into *microfluidic systems* and *organ-on-chips* (Chap. 10). The use of hydrogels in microfluidic chips helps to better recapitulate the in vivo microenvironment providing cells with ECM-like surrounding. On the other hand, microfluidic allows creation of spatiotemporal gradients of bioactive molecules, nutrients, and oxygen in hydrogels in a very small scale [88]. Moreover, microfluidic systems can be created directly from hydrogels [89]. Hydrogels can also be used for *expansion* of various cell types *in stirred tank reactors*—here cells can be encapsulated in hydrogel beads or can grow on the surface of hydrogel microcarriers [90, 91]. Recently, hydrogels were used to enable *3D isolation of MSCs*, resulting in cell material never exposed to plastic adherence in a 2D environment [92]. In the clinic, besides tissue and organ reconstruction by tissue engineering, *injectable hydrogels* are used to protect and support cells for delivery to treatment sites. Injectable hydrogels can be, for example, used for restoring the lost functions of nervous system as drug, liposome and cell delivery systems [93] or wound dressing for skin wounds [94].

5.8   Future Perspectives

The need for more biologically relevant disease models to bridge the gap between in vitro and in vivo conditions has resulted in significant advances being made in the area of 3D cell culture. Hydrogels have been shown to offer great potential in the development of 3D models due to their properties such as high water retention, oxygen and nutrient diffusion and tunability.

The first use of hydrogels for cell culture was reported by Ehrmann and Gey, who in 1956 reconstituted rat tail collagen and used it as substrate for cell growth [95]. Nowadays, hydrogels are used by the scientific community for modeling physiological and pathological tissues, for advanced drug screening and in tissue engineering. In 3D cell culture, the rapidly developing field of 3D bioprinting requires further development of cell promoting, but mechanically stable hydrogels with optimal gelation dynamics, high biocompatibility, and possible biodegradability.

Traditionally animal derived hydrogels have been used but more recently new synthetic, semi-synthetic, and bio-based hydrogels, such as those manufactured from peptides and wood, offer realistic alternatives. When following the journey from cell research in the laboratory to cell therapy at the clinic, it is not possible to utilize a material with animal components. Bio-based, semi-synthetic, and synthetic products offer a clear advantage here, in addition to providing more reproducible manufacturing. Hydrogels produced from recombinant animal proteins also have more potential for clinical applications than ones of animal origin or ones derived from human blood material.

Hydrogels are not limited to plate-based 3D cell culture but have application in organ-on-a-chip models, microfluidic devices, in drug delivery and 3D bioprinting for tissue engineering and regenerative medicine. In all of these applications the ability to support cell health and viability whilst retaining cell morphology and function is paramount. Hydrogels are capable of fulfilling all of these requirements and indeed offer an exciting way forward to bridge the in vitro/in vivo gap and take cell research from the bench to therapy in the clinic.

Take-Home Messages

- Hydrogels are 3D network structures able to imbibe large amounts of water.
- Hydrogels can be isolated from natural animal and non-animal sources, synthesized or a combination of natural and synthetized molecules.
- Unpolymerized hydrogels are liquid and by polymerization become solid (sol-gel transition).
- Hydrogel classification is based on structural composition, origin/source of polymers, responsiveness to stimuli, molecular charge, and crosslinking reversibility.
- Stiffness of the hydrogel plays a role in determining cell behavior and fate and can be adjusted via crosslinking or changing concentration.
- Tunable properties of hydrogels allow creation of various physical/mechanical in vitro microenvironments.
- Hydrogel properties can be characterized by rheology, SEM, confocal microscopy, and AFM.
- Hydrogels can be used to create in vitro gradients.
- Easy removal of the hydrogel is required for cell recovery and downstream analysis but should not damage or affect cell viability.
- Hydrogels are used in standard 3D cell culture, bioprinting, microfluidic devices, organ-on-a-chip models and as carrier material in bioreactors for cell expansion.
- They offer a way to bridge the gap between research and the clinic having use in cell therapy and tissue engineering.

References

1. Patel A, Mequanint K. Hydrogel biomaterials. In: Fazel-Rezai R, editor. Biomedical engineering - frontiers and challenges. Rijeka: InTech; 2011.
2. Drury JL, Mooney DJ. Hydrogels for tissue engineering: scaffold design variables and applications. Biomaterials. 2003;24(24):4337–51.
3. Ruedinger F, et al. Hydrogels for 3D mammalian cell culture: a starting guide for laboratory practice. Appl Microbiol Biotechnol. 2015;99(2):623–36.
4. Dragan ES. Design and applications of interpenetrating polymer network hydrogels. A review. Chem Eng J. 2014;243:572–90.
5. Dinescu S, et al. Collagen-based hydrogels and their applications for tissue engineering and regenerative medicine. In: Mondal M, editor. Cellulose-based superabsorbent hydrogels. Polymers and polymeric composites: a reference series. Cham: Springer; 2018.
6. Ahmed TA, Dare EV, Hincke M. Fibrin: a versatile scaffold for tissue engineering applications. Tissue Eng Part B Rev. 2008;14(2):199–215.
7. Curvello R, Raghuwanshi VS, Garnier G. Engineering nanocellulose hydrogels for biomedical applications. Adv Colloid Interf Sci. 2019;267:47–61.

8. Pepelanova I, et al. Gelatin-methacryloyl (GelMA) hydrogels with defined degree of functionalization as a versatile toolkit for 3D cell culture and extrusion bioprinting. Bioengineering (Basel). 2018;5(3):55.

9. Chen YC, et al. Functional human vascular network generated in photocrosslinkable gelatin methacrylate hydrogels. Adv Funct Mater. 2012;22(10):2027–39.

10. Dikovsky D, Bianco-Peled H, Seliktar D. The effect of structural alterations of PEG-fibrinogen hydrogel scaffolds on 3-D cellular morphology and cellular migration. Biomaterials. 2006;27 (8):1496–506.

11. Seliktar D. Designing cell-compatible hydrogels for biomedical applications. Science. 2012;336 (6085):1124–8.

12. Andersen T, Auk-Emblem P, Dornish M. 3D cell culture in alginate hydrogels. Microarrays. 2015;4(2):133–61.

13. Ekerdt BL, et al. Thermoreversible hyaluronic acid-PNIPAAm hydrogel systems for 3D stem cell culture. Adv Healthc Mater. 2018;7(12):1800225.

14. Hassan W, Dong Y, Wang W. Encapsulation and 3D culture of human adipose-derived stem cells in an in-situ crosslinked hybrid hydrogel composed of PEG-based hyperbranched copolymer and hyaluronic acid. Stem Cell Res Ther. 2013;4(2):32.

15. Chawla K, et al. Biodegradable and biocompatible synthetic saccharide− peptide hydrogels for three-dimensional stem cell culture. Biomacromolecules. 2011;12(3):560–7.

16. Liu Y, Chan-Park MB. Hydrogel based on interpenetrating polymer networks of dextran and gelatin for vascular tissue engineering. Biomaterials. 2009;30(2):196–207.

17. Xiao W, et al. Synthesis and characterization of photocrosslinkable gelatin and silk fibroin interpenetrating polymer network hydrogels. Acta Biomater. 2011;7(6):2384–93.

18. Wisdom K, Chaudhuri O. 3D cell culture in interpenetrating networks of alginate and rBM matrix. In: 3D cell culture. Totowa, NJ: Springer; 2017. p. 29–37.

19. Bhattacharya M, et al. Nanofibrillar cellulose hydrogel promotes three-dimensional liver cell culture. J Control Release. 2012;164(3):291–8.

20. Peppas NA, et al. Hydrogels in biology and medicine: from molecular principles to bionanotechnology. Adv Mater. 2006;18(11):1345–60.

21. Thiele J, et al. 25th anniversary article: designer hydrogels for cell cultures: a materials selection guide. Adv Mater. 2014;26(1):125–47.

22. Tibbitt MW, Anseth KS. Hydrogels as extracellular matrix mimics for 3D cell culture. Biotechnol Bioeng. 2009;103(4):655–63.

23. Thiele J, et al. DNA-functionalized hydrogels for confined membrane-free in vitro transcription/ translation. Lab Chip. 2014;14(15):2651–6.

24. Zhu J. Bioactive modification of poly(ethylene glycol) hydrogels for tissue engineering. Biomaterials. 2010;31(17):4639–56.

25. Almany L, Seliktar D. Biosynthetic hydrogel scaffolds made from fibrinogen and polyethylene glycol for 3D cell cultures. Biomaterials. 2005;26(15):2467–77.

26. Lee KY, Mooney DJ. Alginate: properties and biomedical applications. Prog Polym Sci (Oxf). 2012;37(1):106–26.

27. Klouda L, Mikos AG. Thermoresponsive hydrogels in biomedical applications. Eur J Pharm Biopharm. 2008;68(1):34–45.

28. Ryan DM, Nilsson BL. Self-assembled amino acids and dipeptides as noncovalent hydrogels for tissue engineering. Polym Chem. 2012;3(1):18–33.

29. Heck T, et al. Enzyme-catalyzed protein crosslinking. Appl Microbiol Biotechnol. 2013;97 (2):461–75.

30. Yang G, et al. Enzymatically crosslinked gelatin hydrogel promotes the proliferation of adipose tissue-derived stromal cells. PeerJ. 2016;4:e2497.

31. Mohamed MA, et al. Stimuli-responsive hydrogels for manipulation of cell microenvironment: from chemistry to biofabrication technology. Prog Polym Sci. 2019;98:101147.

32. Qu J, et al. pH-responsive self-healing injectable hydrogel based on N-carboxyethyl chitosan for hepatocellular carcinoma therapy. Acta Biomater. 2017;58:168–80.

33. Nagahama K, Ouchi T, Ohya Y. Temperature-induced hydrogels through self-assembly of cholesterol-substituted star PEG-b-PLLA copolymers: an injectable scaffold for tissue engineering. Adv Funct Mater. 2008;18(8):1220–31.

34. Kloxin AM, Tibbitt MW, Anseth KS. Synthesis of photodegradable hydrogels as dynamically tunable cell culture platforms. Nat Protoc. 2010;5(12):1867.

35. Lim HL, et al. Dynamic electromechanical hydrogel matrices for stem cell culture. Adv Funct Mater. 2011;21(1):55–63.

36. Shen Z, et al. A thermally responsive cationic nanogel-based platform for three-dimensional cell culture and recovery. RSC Adv. 2014;4(55):29146–56.

37. Amosi N, et al. Acidic peptide hydrogel scaffolds enhance calcium phosphate mineral turnover into bone tissue. Acta Biomater. 2012;8(7):2466–75.

38. Green H, et al. RGD-presenting peptides in amphiphilic and anionic β-sheet hydrogels for improved interactions with cells. RSC Adv. 2018;8(18):10072–80.

39. Wang H, Heilshorn SC. Adaptable hydrogel networks with reversible linkages for tissue engineering. Adv Mater. 2015;27(25):3717–36.

40. Discher DE, Janmey P, Wang Y-l. Tissue cells feel and respond to the stiffness of their substrate. Science. 2005;310(5751):1139–43.

41. Rosales AM, et al. Hydrogels with reversible mechanics to probe dynamic cell microenvironments. Angew Chem Int Ed. 2017;56(40):12132–6.

42. Chen Y, et al. Receptor-mediated cell mechanosensing. Mol Biol Cell. 2017;28(23):3134–55.

43. Alakpa EV, et al. Tunable supramolecular hydrogels for selection of lineage-guiding metabolites in stem cell cultures. Chem. 2016;1(2):298–319.

44. Cha C, et al. Decoupled control of stiffness and permeability with a cell-encapsulating poly (ethylene glycol) dimethacrylate hydrogel. Biomaterials. 2010;31(18):4864–71.

45. Ledo AM, et al. Extracellular matrix mechanics regulate transfection and SOX9-directed differentiation of mesenchymal stem cells. Acta Biomater. 2020;110:153–63.

46. Cosgrove BD, et al. N-cadherin adhesive interactions modulate matrix mechanosensing and fate commitment of mesenchymal stem cells. Nat Mater. 2016;15(12):1297–306.

47. Cavo M, et al. A new cell-laden 3D Alginate-Matrigel hydrogel resembles human breast cancer cell malignant morphology, spread and invasion capability observed "in vivo". Sci Rep. 2018;8 (1):1–12.

48. Yosef A, et al. Fibrinogen-based hydrogel modulus and ligand density effects on cell morphogenesis in two-dimensional and three-dimensional cell cultures. Adv Healthc Mater. 2019;8 (13):1801436.

49. Lavrentieva A, et al. Fabrication of stiffness gradients of GelMA hydrogels using a 3D printed micromixer. Macromol Biosci. 2020;20(7):e2000107.

50. Zuidema JM, et al. A protocol for rheological characterization of hydrogels for tissue engineering strategies. J Biomed Mater Res B Appl Biomater. 2014;102(5):1063–73.

51. Shachaf Y, Gonen-Wadmany M, Seliktar D. The biocompatibility of PluronicF127 fibrinogen-based hydrogels. Biomaterials. 2010;31(10):2836–47.

52. Plotkin M, et al. The effect of matrix stiffness of injectable hydrogels on the preservation of cardiac function after a heart attack. Biomaterials. 2014;35(5):1429–38.

53. Kim SW, Bae YH, Okano T. Hydrogels: swelling, drug loading, and release. Pharm Res. 1992;9 (3):283–90.

54. Kirsch M, et al. Gelatin-methacryloyl (GelMA) formulated with human platelet lysate supports mesenchymal stem cell proliferation and differentiation and enhances the hydrogel's mechanical properties. Bioengineering. 2019;6(3):76.
55. Rahman MS, et al. Morphological characterization of hydrogels. In: Mondal MIH, editor. Cellulose-based superabsorbent hydrogels. Cham: Springer International Publishing; 2019. p. 819–63.
56. Shi J, et al. Cell-compatible hydrogels based on a multifunctional crosslinker with tunable stiffness for tissue engineering. J Mater Chem. 2012;22(45):23952–62.
57. Iturri J, Toca-Herrera JL. Characterization of cell scaffolds by atomic force microscopy. Polymers. 2017;9(8):383.
58. Barani A, Bush MB, Lawn BR. Effect of property gradients on enamel fracture in human molar teeth. J Mech Behav Biomed Mater. 2012;15:121–30.
59. Xia T, Liu W, Yang L. A review of gradient stiffness hydrogels used in tissue engineering and regenerative medicine. J Biomed Mater Res A. 2017;105(6):1799–812.
60. Laasanen MS, et al. Biomechanical properties of knee articular cartilage. Biorheology. 2003;40 (1–3):133–40.
61. Tsai AG, Johnson PC, Intaglietta M. Oxygen gradients in the microcirculation. Physiol Rev. 2003;83(3):933–63.
62. Oudin MJ, Weaver VM. Physical and chemical gradients in the tumor microenvironment regulate tumor cell invasion, migration, and metastasis. Cold Spring Harb Symp Quant Biol. 2016;81:189–205.
63. Lewis DM, et al. Intratumoral oxygen gradients mediate sarcoma cell invasion. Proc Natl Acad Sci. 2016;113(33):9292–7.
64. Sansom SN, Livesey FJ. Gradients in the brain: the control of the development of form and function in the cerebral cortex. Cold Spring Harb Perspect Biol. 2009;1(2):a002519.
65. Wartlick O, Kicheva A, Gonzalez-Gaitan M. Morphogen gradient formation. Cold Spring Harb Perspect Biol. 2009;1(3):a001255.
66. Akeson A, et al. Endothelial cell activation in a VEGF-A gradient: relevance to cell fate decisions. Microvasc Res. 2010;80(1):65–74.
67. Pedron S, Becka E, Harley BA. Spatially gradated hydrogel platform as a 3D engineered tumor microenvironment. Adv Mater. 2015;27(9):1567–72.
68. Moeendarbary E, et al. The soft mechanical signature of glial scars in the central nervous system. Nat Commun. 2017;8:14787.
69. Wang L, et al. Hydrogel-based methods for engineering cellular microenvironment with spatio-temporal gradients. Crit Rev Biotechnol. 2016;36(3):553–65.
70. Smith Callahan LA. Gradient material strategies for hydrogel optimization in tissue engineering applications. High-throughput. 2018;7(1):1.
71. Diederich VE, et al. Bioactive polyacrylamide hydrogels with gradients in mechanical stiffness. Biotechnol Bioeng. 2013;110(5):1508–19.
72. Jeon O, et al. Biochemical and physical signal gradients in hydrogels to control stem cell behavior. Adv Mater. 2013;25(44):6366–72.
73. Burdick JA, Khademhosseini A, Langer R. Fabrication of gradient hydrogels using a microfluidics/photopolymerization process. Langmuir. 2004;20(13):5153–6.
74. Zaari N, et al. Photopolymerization in microfluidic gradient generators: microscale control of substrate compliance to manipulate cell response. Adv Mater. 2004;16(23–24):2133–7.
75. Sundararaghavan HG, et al. Neurite growth in 3D collagen gels with gradients of mechanical properties. Biotechnol Bioeng. 2009;102(2):632–43.
76. Orsi G, et al. A new 3D concentration gradient maker and its application in building hydrogels with a 3D stiffness gradient. J Tissue Eng Regen Med. 2017;11(1):256–64.

77. Lee D, et al. Fabrication of hydrogels with a stiffness gradient using limited mixing in the Hele-Shaw geometry. Exp Mech. 2019;59(9):1249–59.
78. Sunyer R, et al. Fabrication of hydrogels with steep stiffness gradients for studying cell mechanical response. PLoS One. 2012;7(10):e46107.
79. Nemir S, Hayenga HN, West JL. PEGDA hydrogels with patterned elasticity: novel tools for the study of cell response to substrate rigidity. Biotechnol Bioeng. 2010;105(3):636–44.
80. Zervantonakis IK, et al. Three-dimensional microfluidic model for tumor cell intravasation and endothelial barrier function. Proc Natl Acad Sci U S A. 2012;109(34):13515–20.
81. Vega SL, et al. Combinatorial hydrogels with biochemical gradients for screening 3D cellular microenvironments. Nat Commun. 2018;9(1):614.
82. Peerani E, Candido JB, Loessner D. Cell recovery of hydrogel-encapsulated cells for molecular analysis. In: Theranostics. New York: Springer; 2019. p. 3–21.
83. Shu XZ, et al. Disulfide cross-linked hyaluronan hydrogels. Biomacromolecules. 2002;3(6):1304–11.
84. Raza A, Lin C-C. Generation and recovery of β-cell spheroids from step-growth PEG-peptide hydrogels. J Vis Exp. 2012;70:e50081.
85. Zhang J, Skardal A, Prestwich GD. Engineered extracellular matrices with cleavable crosslinkers for cell expansion and easy cell recovery. Biomaterials. 2008;29(34):4521–31.
86. Lou Y-R, et al. The use of nanofibrillar cellulose hydrogel as a flexible three-dimensional model to culture human pluripotent stem cells. Stem Cells Dev. 2014;23(4):380–92.
87. Bidarra SJ, Barrias CC. 3D culture of mesenchymal stem cells in alginate hydrogels. In: Stem cell niche. New York: Springer; 2018. p. 165–80.
88. Lee SH, et al. Hydrogel-based three-dimensional cell culture for organ-on-a-chip applications. Biotechnol Prog. 2017;33(3):580–9.
89. Paguirigan A, Beebe D. Gelatin based microfluidic devices for cell culture. Lab Chip. 2006;6(3):407–13.
90. Huang L, et al. Biopolymer-based microcarriers for three-dimensional cell culture and engineered tissue formation. Int J Mol Sci. 2020;21(5):1895.
91. Kumar A, Starly B. Large scale industrialized cell expansion: producing the critical raw material for biofabrication processes. Biofabrication. 2015;7(4):044103.
92. Egger D, et al. From 3D to 3D: isolation of mesenchymal stem/stromal cells into a three-dimensional human platelet lysate matrix. Stem Cell Res Ther. 2019;10(1):248.
93. Pakulska MM, Ballios BG, Shoichet MS. Injectable hydrogels for central nervous system therapy. Biomed Mater. 2012;7(2):024101.
94. Qu J, et al. Antibacterial adhesive injectable hydrogels with rapid self-healing, extensibility and compressibility as wound dressing for joints skin wound healing. Biomaterials. 2018;183:185–99.
95. Ehrmann RL, Gey GO. The growth of cells on a transparent gel of reconstituted rat-tail collagen. J Natl Cancer Inst. 1956;16(6):1375–403.

Vascularization in 3D Cell Culture

6

M. Markou, D. Kouroupis, T. Fotsis, E. Bagli, and C. Murphy

Contents

M. Markou · E. Bagli (✉) · C. Murphy (✉)
Foundation for Research & Technology-Hellas (FORTH), Department of Biomedical Research,
Institute of Molecular Biology and Biotechnology (IMBB), Ioannina, Greece
e-mail: mmarkou92@gmail.com; elenibgl@hotmail.com; carol_murphy@imbb.forth.gr

D. Kouroupis
Department of Orthopaedics, Division of Sports Medicine, Diabetes Cell Transplant Center,
University of Miami, Miller School of Medicine, Miami, FL, USA
e-mail: dxk504@med.miami.edu

T. Fotsis
Foundation for Research & Technology-Hellas (FORTH), Department of Biomedical Research,
Institute of Molecular Biology and Biotechnology (IMBB), Ioannina, Greece

Laboratory of Biological Chemistry, Medical School, University of Ioannina, Ioannina, Greece
e-mail: thfotsis@uoi.gr

© Springer Nature Switzerland AG 2021
C. Kasper et al. (eds.), *Basic Concepts on 3D Cell Culture*, Learning Materials in
Biosciences,
https://doi.org/10.1007/978-3-030-66749-8_6

Abbreviations

2D	Two-dimensional
3D	Three-dimensional
BMP-4	Bone morphogenetic protein-4
CDM	Contractile differentiation medium
cSMCs	Contractile smooth muscle cells
DMSO	Dimethyl sulfoxide
ECGS	Endothelial cell growth supplement
ECM	Extracellular matrix
ECs	Endothelial cells
EGF	Epidermal growth factor
FCS	Fetal calf serum
FGF2	Fibroblast growth factor 2
GSK-3	Glycogen synthase kinase-3
hESCs	Human embryonic stem cells
hiPSCs	Human induced pluripotent stem cells
hPSCs	Human pluripotent stem cells
HSA	Human serum albumin
HUVECs	Human umbilical vein endothelial cells
IGF-I	Insulin-like growth factor
MCs	Mural cells
PBS	Phosphate buffer saline
PCs	Pericytes
RBCs	Red blood cells
SDM	Synthetic differentiation medium
SMCs	Smooth muscle cells
sSMCs	Synthetic smooth muscle cells
VEGF-A	Vascular endothelial growth factor-A

What You Will Learn in This Chapter

The two-dimensional (2D) monoculture does not recapitulate the three-dimensional (3D) environment of in vivo vasculature. Endothelial cell (ECs) spheroids have been introduced as a 3D in vitro model exhibiting angiogenic responses and sprouting behavior in vivo. However, mural cells (MCs), including the phenotypic range from pericytes to smooth muscle cells (SMCs), play an essential organotypic role in vascular homeostasis (vascular remodeling/stabilization) and disease. Therefore, the development of 3D cell structures (small-scale vascular organoids) containing both ECs and MCs subtypes provide enhanced cell–cell interactions that closely mimic the natural/physiological vascular microenvironment with beneficial effects on cell survival, phenotypic stability, and function. Due to limitations of isolation/ expansion of primary MCs as well as their phenotypic plasticity during in vitro culture, human pluripotent stem cells (hPSCs) (human induced pluripotent stem cells-hiPSCs and human embryonic stem cells-hESCs) are a source for the generation of defined MC populations.

We generated a flexible, small-scale 3D organoid-like platform consisting of hPSC-SMCs/ECs, which undergo self-assembly into a segregated 3D structure characterized by a multicellular spheroidal SMC core and an outer EC layer. The structure can be regarded as an inside-out assembly of a resting vessel wall. When these vascular organoids are implanted in 3D extracellular matrices, they are superior to sole ECs regarding the development of a mature vascular network.

Questions

- Which are the vascular cells and why are both necessary for studying angiogenesis in vitro?
- Which are the advantages and the limitations of using hPSCs as a source of vascular cells?
- Why are 3D cell cultures superior to 2D?
- What are the advantages and limitations of using a 3D co-culture system?

Overview of the Procedure

A general outline of the procedure and the duration of all individual steps are presented as a flow diagram in Fig. 6.1.

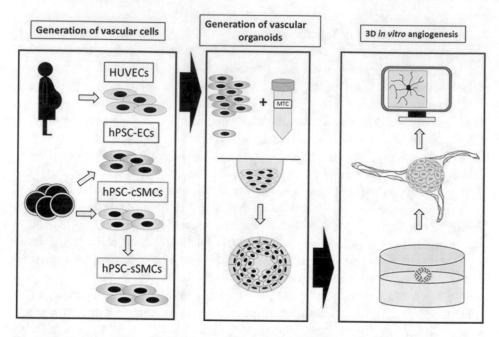

Fig. 6.1 Overview of the procedure. Vascular cells (ECs and SMCs) differentiated from hPSCs or ECs isolated from umbilical vein are co-cultured with methylcellulose (MTC) as hanging drops in order to create vascular organoids. After their phenotypic analysis, vascular organoids are implanted in 3D matrices generating a 3D vascular network in vitro

Procedure
1. Differentiation of hPSCs to vascular cells.
2. Generation of the vascular organoids.
    a. Preparation of methylcellulose.
    b. Fabrication of the vascular organoids.
    c. Phenotypic analysis of the vascular organoids.
3. Incorporation of vascular organoids in 3D matrices (generation of 3D vascular network in vitro).

Introduction

Regenerative Medicine focuses on repair, replacement, or regeneration of cells, tissues, or organs to restore impaired function resulting from congenital defects, disease, and trauma [1]. A major requirement for viability and function of the implantable construct is the availability of blood vessels to support its in vivo growth. Vascularization remains a critical obstacle in engineering thicker, metabolically demanding organs, such as heart muscle, brain, and liver as regenerating tissue over 100–200 μm exceeds the capacity of nutrient supply and waste removal by diffusion, and requires a vascular network [2, 3]. It takes several weeks for a scaffold to become fully vascularized in vivo [4], as host vessel ingrowth proceeds slowly, at a rate of less than several tenths of a micrometer per day and without a rapid and high level of vascularization of the transplanted grafts, the majority of cells fail to survive the early post transplantation phase. Therefore, the development of strategies that enhance the angiogenic process represents one of the major research topics in the field of tissue engineering.

The current strategies for generating 3D vascularized construct can be categorized as cell-based, angiogenic factor-based, or scaffold-based approaches. Scaffold-based approaches include the construction of tissues with decellularized grafts, sacrificial scaffolds, spatial micropatterning, biomimetic scaffolds using vascular corrosion casting, and 3D printing techniques (reviewed in [5]). The current classical cell-based approach including the isolation, expansion, and seeding of endothelial cells (ECs) onto a suitable scaffold before in vivo implantation [6], leads to the generation of an immature vascular network in vivo, the ineffective integration of this network into the host vasculature and subsequently to the regression of the vessels within a few days [7, 8]. Moreover, the generated capillaries are leaky and unable to properly control permeability, contributing to tissue edema [9, 10]. Therefore, a particular challenge for the tissue engineering community is to induce vascularization of ischemic tissues with blood vessels that are functionally normal. To promote the maturation and stability of nascent vasculatures, ECs must functionally interact with MCs.

In the last few years, it has become apparent that when cells are cultured in 3D they adhere to each other via ECM and form natural cell–cell contacts, which transmit physiological information regulating cell growth, migration, differentiation, and survival [11, 12] resembling the native environment. In this context, generation of 3D cell structures (small-scale vascular organoids) containing both ECs and SMCs subtypes, unlike traditional 2D monolayer co-cultures, provide enhanced cell–cell interactions that closely mimic the natural/physiological vascular microenvironment with beneficial effects on cell survival, phenotypic stability and function [11, 12]. Especially for MCs, given their heterogeneity and spatiotemporal variation in protein expression patterns, the development of such 3D cell platforms could preserve their phenotypic signature. In 1998 Korff et al. [13] introduced 3D EC spheroids as an in vitro model exhibiting angiogenic responses and sprouting behavior in vivo. Since then, multicellular spheroids have become a common 3D cell culture system, generated either from one or many cell types for multiple applications (reviewed in Lascke et al. [14]). Utilizing this simple approach of fabricating 3D vascular organoids of MC subtypes and ECs we have shown that they are superior to mixed monocells and sole ECs regarding the development of a mature vasculature in vivo and furthermore, they are ready to use for various tissue engineering applications [15]. Their generation could also be the first step in designing more complex 3D tissue engineering constructs (by including organ specific cells, growth factors, scaffolds). Finally, these vascular organoids are a defined in vitro model for studying the paracrine interactions between ECs and MC subtypes that regulate vessel assembly, phenotype modulation, maturation, maintenance, and vessel destabilization in a way that mimics the physiological assembly of the normal vasculature and therefore might serve as a platform for disease modeling and drug development, including estimations of compound preclinical toxicity and potential metabolic liability.

6.1   Generation of Vascular Cells-2D Culture

Blood vessels are composed of ECs and MCs. ECs form a thick layer, called endothelium, in the inner surface of the vessels and are in contact with the blood. The endothelium acts as a selective barrier for the passage of constituents and gases between the lumen and surrounding tissues.

MCs are primarily responsible for stabilization, inhibition of regression, contraction of the vessel as well as production and deposition of extracellular matrix (ECM) proteins [16, 17]. Interactions between MCs and ECs are critical in the process of vascular development [18–20]. MCs are composed of SMC surrounding larger vessels, such as arteries and veins, and pericytes (PCs) typically surrounding smaller microvessels and capillaries. However, according to recent literature, MCs consist of a phenotypic spectrum with PCs at the one end and SMCs at the other [21]. As a result, MCs exhibit overlapping marker expression and cannot be distinguished by one marker alone. Regarding SMCs, two distinct phenotypes have been identified: synthetic and contractile [22, 23]. Both participate in neovascularization, but synthetic vSMCs predominate in the embryo and in diseased or injured adult vessels while contractile vSMCs predominate in healthy adult vessels.

The development of in vitro angiogenesis/vasculogenesis models that recapitulate the physiological assembly of the normal vasculature is critical to study the paracrine interactions between ECs and SMC subtypes, which are responsible for the vessel assembly, phenotype modulation, maturation, maintenance, and vessel destabilization. However, the cell source and culture conditions of the vascular cells required for the development of these models can be challenging. The "traditional " source of human cells used in the laboratory is the human body from which they can be isolated (primary cells). ECs, for instance, are widely isolated from human umbilical vein, but concerning MCs there are a lot of restrictions regarding the organs from which they can be isolated, the number of cells that can be obtained as well as the expansion and the phenotypic stability during their culture [21].

Human pluripotent stem cells (hPSCs), including induced PSCs (iPSCs) and embryonic stem cells (ESCs), can differentiate into the three germ layers. They have an unlimited ability to self-renew, making them easy to expand and, despite limitations of the current differentiation procedures [24], represent an unlimited source of cells for therapeutic use. The generation of iPSCs, although laborious and expensive, overcomes the ethical problems associated with the clinical use of stem cells and provides the possibility of using autologous cells [25].

The protocols regarding isolation of primary ECs human umbilical vein (HUVECs), differentiation of hPSCs to ECs (hPSC-ECs) differentiation of hPSC to SMC subtypes (contractile SMCs-hPSC-cSMCs and synthetic SMCs-hPSC-sSMCs) are described below.

6.1.1 Isolation of Primary ECs from Umbilical Vein

1. The blood from cords are rinsed in the Biosafety Cabinet (class II) and both ends are cut off the cord with a scalpel.
2. Using a plastic syringe (20 mL) the vein is washed with PBS (the vein is the largest opening; the two smaller ones are arteries).
3. Two three-way stopcocks (Helm pharmaceuticals GMBH) are inserted into the vein, one on each end, and they are tightly immobilized with silk surgical suture (Medipac, #116).
4. Vein is further washed as in step 2 with PBS, so that all red blood cells (RBCs) are removed.
5. The vein is then filled with collagenase type 1 solution (0.2% diluted in PBS) (Worthington, LS004196) with a syringe. When it leaks out of the other extremity, the open end is clamped and collagenase is infused until there is moderate distention of the vein.
6. The cord (with stopcocks and collagenase) is incubated in a beaker filled with preheated PBS at 37 °C in a waterbath for 13 min.
7. After incubation, the cord is removed from the beaker and its content is collected in a 50 mL tube.
8. Then, the vein is flushed with 15–25 mL M199 medium (Gibco, 31,150,022) supplemented with 10%FCS (Gibco, 10,270–106) and the wash is added to the same tube from step 7.
9. The content is centrifuged at 1200 rpm for 10 min at RT.
10. The supernatant is carefully discarded.
11. The pellet of cells is suspended in "complete M199 medium."*.
12. The isolated cells of 3–4 umbilical veins are transferred to a collagen coated 10 cm plate**.
13. Next day the medium is changed in order to remove dead cells and RBCs (a wash with PBS can be performed carefully).
14. After 2 days cells should be confluent and ready for passaging.
15. Medium is removed, cells are washed once with PBS and 1 mL of 0.05% Trypsin-EDTA (Gibco, 25,300–054) are added. The trypsin is immediately removed and the cells incubated at 37 °C for about 1 min. Complete M199 medium is added, cells detached and plated onto collagen coated plates in a ratio of 1:3. The phenotype is stable for at least 3 passages (based on the presence of ECs characteristic factors, such as von *Willebrand* factor).

*M199 basal medium (Gibco, 31,150,022) supplemented with 20% FCS, 47 μg/mL ECGS, (Sigma, E2759), 4.7u/mL heparin (Sigma, H3149), and 1% penicillin-streptomycin.

**Preparation of Collagen coated plates: 50 μg/mL collagen I (Corning, 354,236) is diluted using 0.02 N acetic acid (Sigma, 33,209-M) in tissue culture water (HyClone, SH30529.02) and filtered (0.22 μm pore size). 5–6 mL of this solution is added to the 10 cm plate to cover all the surface. After 30 min incubation at 37 °C, the plate is washed twice with PBS and it is ready to use.

Note: Washes with PBS are mandatory because collagen solution is acidic. Plates once washed with PBS must be used immediately in order to avoid drying out.

Tip: In order to create a pool of ECs, at least three umbilical veins are used each time for ECs isolation.

6.1.2   Generation of ECs from hPSCs (hPSC-ECs)

1. hPSCs and more specifically hESCs (H1 cells from WiCell, Madison, WI, USA) or hiPSCs (we use hiPSCs generated in our lab from human fibroblasts [26]) are cultured as colonies on six-well tissue culture plates coated with hESC-qualified Matrigel (Corning, 354,277) in mTeSR[1] medium (STEMCELL Technologies, 05850).
2. hESCs/hiPSCs colonies are first dissociated into small clumps (using 1 mg/mL dispase) and then replated onto Matrigel coated six-well plates in mTesR[1] medium as per normal routine passaging*.
3. After 48 h (Day 0), mTeSR[1] medium is changed to differentiation medium (APEL), which is synthesized as previously described [27] ** and supplemented with 5 μM GSK-3 inhibitor (Selleckchem, CHIR99021, stock solution prepared in DMSO (Sigma, D2438)) for 24 h.
4. Day 1, the medium is replaced with APEL medium supplemented with 25 ng/mL BMP-4 (Life Technologies, PHC9534, stock solution prepared in 0.1% HSA solution, Vitrolife 10,064) for 48 h.
5. Day 3, medium is changed to APEL medium supplemented with 80 ng/mL VEGF-A (Immunotools, 11,343,663, stock solution prepared in 0.1% HSA) for 48 h.
6. Day 5, CD34+ ECs (around 40% of the cells) are isolated using the EasySep Human CD34 Positive Selection Kit (STEMCELL Technologies, 18,056) according to the manufacturer's instructions and cultured on fibronectin coated 24 well plates (100,000 cells/well)*** in EGM-2 medium (Lonza, CC-3162).
7. After 1–2 days, plates should be confluent with phenotypical and functional ECs (Fig. 6.2) [28].

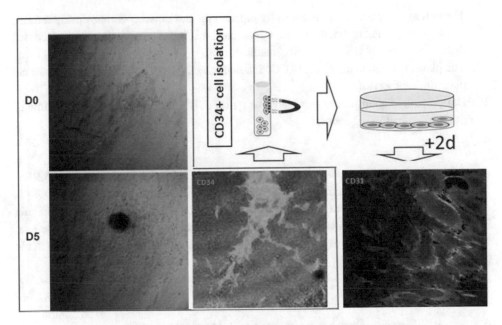

Fig. 6.2 Generation of hPSC-ECs. Representative photographs of hPSCs ready to begin the differentiation-D0 (upper image) and on D5 of the differentiation procedure (lower, left image) are shown. Cells were imaged on Zeiss Axiovert 100 using a 5X objective. CD34+ cells on D5 are shown. CD34 positive cells are shown in green before their isolation (using anti-CD34-FITC Ab, direct immunofluorescence, middle image) and two days after their isolation (using anti-CD31 Ab, indirect immunofluorescence, right image). Images were taken on Leica TCS SP5 confocal microscope

1. *When the edges of individual colonies are close together and colonies are compact with dark dense centers, they are ready for passaging.
2. Prior to passaging, an aliquot of frozen matrigel is diluted in 25 mL of DMEM/F12 (HyClone, SH30023.01) according to manufacturer's instructions and used to coat new 6-well dishes (1 mL/well). Matrigel is allowed to solidify for 1 h at RT and the dishes are subsequently washed once with 1 mL DMEM/F12.
3. Medium is aspirated from the well containing the hPSCs and cells are washed with 2 mL of DMEM/F-12.
4. 1 mL of dispase (1 U/mL) (Invitrogen, 17,105–041, diluted in DMEM/F12) is added and incubated at 37 °C for 1 min.
5. The dispase is aspirated, and each well is gently washed twice with 2 mL of DMEM/F-12.
6. 2 mL of mTeSR™1 are added and colonies are gently detached by scraping with a cell scraper (Corning, 3010).
7. The detached cell aggregates are transferred to a 15 mL conical tube and a wash with 2 mL additional mTeSR™ is repeated to collect any remaining aggregates.

8. The cell aggregate mixture is carefully pipetted up and down 2–3 times using a 5 mL serological pipette to break up the aggregates and is transferred to matrigel coated dishes in a ratio of 1:6, in mTeSR^1 medium.
9. The plate is placed in a 37 °C/5% CO^2 incubator for 24 h being careful not to move the plate during this time.
10. Daily medium changes are performed using mTeSR^1 until the next passaging time or the differentiation procedure.

*Alternatively, APEL medium can be purchased (STEMCELL Technologies, 05210).

**Preparation of fibronectin coated plates: human Fibronectin (Corning, 354,008) is diluted in PBS and added to the plate to cover the surface (5 µg/cm^2). After 30 min incubation at RT the plate is washed once with tissue culture water and it is ready to use.

Note: Plates once washed with PBS must be used immediately in order to avoid drying out.

6.1.3   Generation of Contractile SMCs (cSMCs) from hPSCs

1. hPSCs (hESCs/ hiPSCs) are grown as colonies on six-well tissue culture plates coated with hESC-qualified Matrigel in mTeSR^1 medium.
2. After 2 days of a normal routine passaging (Sect. 6.1.2), cells are washed once with plain DMEM-F12 (HyClone, SH30023.01) and their culture medium is changed to contractile differentiation medium (CDM)*, which is changed every day for 9 days.
3. On day 9, cells are detached enzymatically using 0.05% Trypsin-EDTA (Gibco, 25,300–054), replated (ratio 1:1), without sorting, onto gelatin coated plates** and cultured in CDM. Once confluent, almost all the cells acquire the desired phenotype (hPSC-cSMCs) and can be used (Fig. 6.3). The phenotype and function of hPSCs derived cSMCs (hESC/hiPSC-cSMCs) is confirmed using immunofluorescence, flow cytometry, and functional assays [15].
4. hPSC-cSMCs are routinely cultured on gelatin coated plates in CDM, which is changed every second day. Their phenotype is stable for at least eight passages.

*CDM: Basal medium (Lonza, PT-3273) supplemented with 2.5% FCS and glutamax (1x, Gibco, 35,050).

**Preparation of gelatin coated plates: Gelatin solution (0.1% Gelatin-Millipore, ES-006-B) is added to the plate to cover all the surface. After 30 min incubation at 37 °C the plate is washed once with PBS and is ready to use.

NOTE: Plates once washed with PBS must be used immediately in order to avoid drying out.

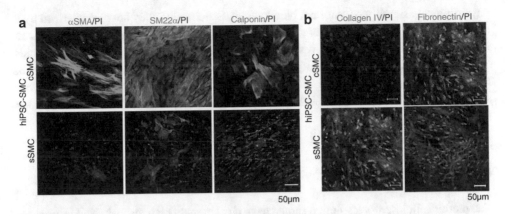

Fig. 6.3 Phenotypic characterization of hPSC-SMC subtypes. A. Immunofluorescence analysis performed on hiPSC-cSMCs and hiPSC-sSMCs. Green fluorescence indicates cells positive for αSMA (left column), SM22α (middle column), and Calponin (right column), whereas red indicates nuclei (PI stain). Images were taken on Leica TCS SP5 confocal microscope. Scale bar, 50 μm. B. Extracellular matrix deposition. Immunofluorescence analysis performed on hiPSC-cSMCs and hiPSC-sSMCs. Green fluorescence indicates Collagen (left column), and Fibronectin (right column), whereas red fluorescence indicates nuclei (PI stain). Images were taken on Leica TCS SP5 confocal microscope. Scale bar, 50 μm. Figure adapted from Markou et al. Frontiers in bioengineering and biotechnology 2020;8:278 [14]

6.1.4 Generation of Synthetic SMCs (sSMCs) from hPSCs

1. hESC/hiPSC-cSMCs cultured on gelatin coated plates (generated as described in Sect. 6.1.3) are washed once with PBS and then synthetic differentiation medium (SDM)* is added.
2. After 2d the synthetic phenotype and function of hPSC-SMCs is acquired (hESC/hiPSC-sSMCs) (Fig. 6.3) [15].

*SDM medium: Basal medium (ScienCell, 1201-b) supplemented with 2% FCS, and a combination of growth factors (ScienCell, 1252) including EGF, 2 ng/mL, FGF2, 2 ng/mL, and IGF-I, 2 ng/mL.

NOTE: The presence of FGF2 is critical for obtaining the synthetic phenotype, therefore SDM can be substituted with CDM supplemented only with FGF2 (2 ng/mL) [15].

6.2 Generation and Characterization of 3D Vascular Organoids

We efficiently generated 3D SMC-EC spheroids (vascular organoids) using the two hPSC-SMC subtypes (cSMCs and sSMCs) and ECs. Randomly mixed hPSC-cSMC/EC or hPSC-sSMC/EC in a fixed ratio of 1/9, the average ratio of MC:EC that exists in the human vasculature, undergo self-assembly into a segregated 3D structure representing the physiological assembly of a normal blood vessel. Specifically, they are characterized by a multicellular spheroidal SMC core and an outer EC layer, which can be regarded as an inside-out assembly of a resting vessel wall (Fig. 6.4). Analysis of the vascular organoids reveals preservation of their phenotypic signatures, while their implantation in matrigel or hydrogels composed of defined ECM components leads to increased capillary network sprouting, which is characterized by SMC-EC co-alignment within the generated sprouts [15].

6.2.1 Preparation of Methylcellulose Solution

The methylcellulose stock solution should have an extremely high viscosity and is prepared as previously described [13]. Specifically:

1. Methylcellulose powder (6 g) (Sigma, M0512) is autoclaved in a 500 mL flask containing a magnetic stirrer (the methylcellulose powder is resistant to this procedure).
2. The autoclaved methylcellulose is dissolved in 250 mL of preheated (60 °C) basal medium (DMEM, Sigma D6046) for 20 min (using the magnetic stirrer).
3. Thereafter, a further 250 mL basal medium (RT) is added to reach a final volume of 500 mL and the solution is mixed for 1–2 h at 4 °C.
4. The final stock solution is aliquoted and cleared by centrifugation (5000 g, 2 h, RT). Only the clear highly viscous supernatant should be used for the generation of the organoids (about 90–95% of the stock solution) (stock solution should be kept at 4 °C).

6.2.2 Preparation of the Cells

1. Both cell types, ECs (HUVECs or hPSC-ECs) and hPSC-SMCs (cSMCs or sSMCS), once confluent are detached enzymatically from the plate using 0.05% Trypsin-EDTA (as for regular passage), centrifuged (1000 g for 5 min, RT) and then resuspended in a small volume of EGM2 medium to estimate cell number.
2. *Cells can be labeled with general membrane dyes (PKH26-red fluorescence and PKH67-green fluorescence) according to manufacturer's instructions in order to be distinguished from each other during the assays.
3. EGM-2 medium is added in order to acquire the desired cell concentration**.

*This is an optional step. Alternatively, cells can be infected with lentiviruses expressing fluorescent proteins (GFP/RFP). NOTE: Cells stained with membrane dyes, retain a uniform membrane staining only for a few days, while cells infected using lentiviruses are permanently labeled use to viral integration).

**Each vascular organoid consists of 1000 cells in a ratio of 9:1 ECs-SMCs (ECs-cSMCs or ECs-sSMCs) in a volume of 10 μL (2 μL methylcellulose stock solution,

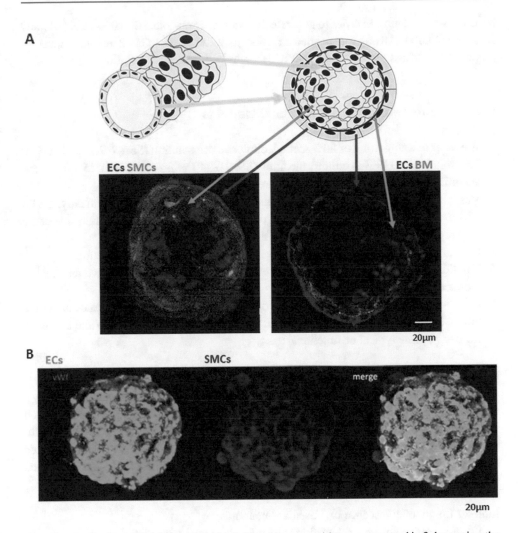

Fig. 6.4 Generation of vascular organoids. A. Vascular organoids were generated in 2 days using the hanging drop method. ECs form a layer on the surface of the organoids (illustration). Immunofluorescence analysis was performed on vascular organoids consisting of hiPSC-SMC/EC. Green fluorescence indicates SMCs expressing αSMA or proteins of basement membrane (Collagen IV), whereas red indicates ECs (von Willebrand staining) and blue indicate nuclei (Draq5 staining). Images were taken on Leica TCS SP5 confocal microscope. Scale bar, 20 μm. B. Representative photo of a vascular organoid witch consist of 1000 cells in a ratio hESC-SMC/EC 1/9. Green fluorescence indicates hESC-SMCs (prelabeled with PKH67) and red immunofluorescence indicates ECs (prelabeled with PKH26, left column), and the merge is shown in the right column. Images were taken on Leica TCS SP5 confocal microscope. Scale bar, 20 μm

8 µL cells in medium). In order to generate 200 vascular organoids, 200,000 cells (180,000 ECs and 20,000 sMCs) resuspended in 1600 µL (200 × 8) EGM-2 medium must be prepared and mixed with 400 µL (200 × 2) methylcellulose stock solution.

6.2.3 Generation of the Vascular Organoids

1. A master mix consisting of 20% methylcellulose stock solution (Sect. 6.2.1) and 80% of cell solution in EGM-2 culture medium (Sect. 6.2.2) is prepared in a 15 mL falcon according to the final number of vascular organoids required.
2. Vascular organoids are generated using the hanging drop technique. Specifically, the lid of a 10 mm plate is inverted and 10 µL drops of the mix are added to the underside of the lid keeping a distance between them so that they do not mix.
3. PBS is added to the dish to protect the vascular organoids from drying out.
4. The lid is placed onto the PBS-filled bottom plate, which is then transferred and kept in the incubator (37 °C, 5%CO_2).
5. After 48 h, organoids are generated and observed as spherical cell aggregates under the microscope. They are then collected with caution using a 200 µl pipet tip and are ready for use in phenotypical or functional assays.

6.2.3.1 Phenotypic Characterization of the Vascular Organoids

ECs and SMC subtypes undergo a self-assembly into vascular organoids characterized by a multicellular spheroidal SMC core and an outer EC monolayer. The distribution of the cell types in the vascular organoids, their phenotype and interactions, as well as the extracellular matrix deposition can be evaluated using indirect immunofluorescence. The generated SMC-EC mixed spheroids (vascular organoids) preserve the phenotypic and functional signatures of the two SMCs subtypes [14].

Indirect Immunofluorescence of Vascular Organoids
1. Vascular organoids are collected in a 15 mL falcon tube with PBS.
2. They are centrifuged at 600 g for 1 min at RT and the supernatant is discarded.
3. A small amount of PBS is added, the pellet is loosened and transferred into a 1.5 mL eppendorf tube.
4. 500 µL of Paraformaldehyde (3.7% in PBS, Merck, 104,005) is added for fixation and incubated for 1 h, RT.
5. Fixed organoids are then permeabilized with 0.2% Triton-X/0.9% gelatin solution for 1 h, and subsequently incubated with 0.5% Triton-X/0.9% gelatin solution for 15 min.
6. Primary antibody (Table 6.1) solution is added and incubated overnight at 4 °C under gently agitation.

Table 6.1 Primary and secondary antibodies used in the study

Antibody	Company	Cat. No.	Working concentration
αSMA	DAKO	M0851	0.71 µg/mL
Calponin	DAKO	M3556	0.3 µg/mL
SM22α	ABCAM	Ab14106	2.5 µg/mL
Collagen IV	DSHB	M3F7-s	2 µg/mL
Fibronectin	DSHB	P1H11-S	0.92 µg/mL
CD31	DSHB	P2B1-s	2 µg/mL
VEGFRII	Cell signaling	2479S	1/100 dilution
Von Willebrand factor	DAKO	A0082	13.6 µg/mL
Alexa Fluor® 488 AffiniPure donkey anti-mouse IgG (H + L)	Jackson ImmunoResearch Laboratories	715–545-151	5 µg/mL
Alexa Fluor® 488 AffiniPure donkey anti-rabbit IgG (H + L)	Jackson ImmunoResearch Laboratories	711–545-152	5 µg/mL
Alexa Fluor® 594 AffiniPure donkey anti-mouse IgG (H + L)	Jackson ImmunoResearch Laboratories	715–585-151	5 µg/mL
Rhodamine (TRITC) AffiniPure donkey anti-rabbit IgG (H + L)	Jackson ImmunoResearch Laboratories	711–025-152	7.5 µg/mL

7. Next day, the vascular organoids are washed 5 times with 0.2% Triton-X/0.9% gelatin solution in PBS and are incubated with secondary antibody (Table 6.1) solution for 1 h at RT.
8. After rinsing 5 times with 0.2% Triton-X, they are incubated with Draq5 for nuclear staining (ThermoFisher Scientific, 62,254) for 10 min and washed with PBS.
9. Images are taken on a Leica TCS SP5 confocal microscope using HCX PL APO CS 40x 1.25 oil objective.

Note 1: In all procedures vascular organoids are in suspension. In order to incubate them in a new solution they are centrifuged at 600 g for 1 min, the supernatant is carefully aspirated and the new solution is added. Alternatively, vascular organoids can be allowed to sink to the bottom of the eppendorf for 5 min without any agitation, due to gravity.

Note 2: Quite a few organoids are lost in the process, therefore it is best to start with a number larger than that required.

6.3   Angiogenic Potential of Vascular Organoids In Vitro

The vascular organoids serve as starting focal points giving rise to sprouts where SMC and EC co-align and exhibit the potential to give rise to a durable (compared to monocells) 3D vascular network, when they are implanted in matrigel in vitro [15]. They also exhibit a similar sprouting profile when implanted in hydrogels consisting of defined ECM components (collagen I, fibronectin, and fibrinogen), which are devoid of any incorporated growth factors and therefore, present an ideal environment for studying the molecular mechanisms underlying the angiogenic process. The morphometric characteristics of the networks generated using vascular organoids consisting of different SMC subtypes/ECs implanted in various hydrogels can be easily evaluated using ImageJ software.

6.3.1   In Vitro Sprouting Assay Using Matrigel

6.3.1.1 Preparation of Matrigel
1. The day before seeding vascular organoids, Matrigel (Corning, 354,234) is placed on ice, at 4 °C. The gel can slowly thaw overnight.

6.3.1.2 Sprouting Assay
1. Matrigel (10 μL) is applied to each inner well of a μ-Slide Angiogenesis plate (ibidi, 81,506). Precooled pipet tips (4 °C) must be used for pipetting the gel.
2. A petri dish with water-soaked paper towels for use as an extra humidity chamber is prepared. The μ-Slide is placed in the petri dish and the lid is closed.
3. The whole assembly is placed into the incubator (37 °C, 5%CO_2) and matrigel is allowed to polymerize for 60 min.
4. One vascular organoid in 40 μL EGM-2 medium is added to each well containing the polymerized matrigel using a 200 μL pipet. The μ-Slide is then returned to the incubator.
5. Medium is changed every 2 days and the sprouting is observed daily and images are taken using Leica TCS SP5 confocal microscope.
6. Quantification is performed using ImageJ software. (Fig. 6.5).

Note: Avoid bubbles

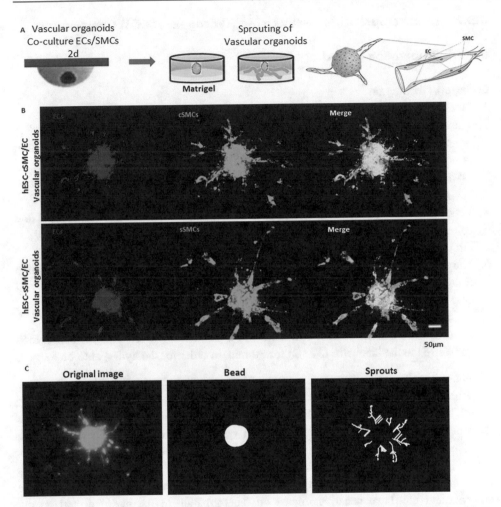

Fig. 6.5 In vitro angiogenesis/sprouting assay on matrigel. A. hPSC-SMC subtype/EC vascular organoids generated using the hanging drop method were added on matrigel and allowed to develop sprouts (illustration). B. Representative images of hESC-cSMC/EC and hESC-sSMC/EC vascular organoids 3d after their addition to matrigel. Red immunofluorescence indicates ECs (prelabeled with PKH26, left panel), green fluorescence indicates hESC-cSMCs and hESC-sSMCs (prelabeled with PKH67, middle panel), and the merge is shown in the right panel. Images were taken on a Leica TCS SP5 confocal microscope. Scale bar, 50 μm. Figure adapted from Markou et al. Frontiers in bioengineering and biotechnology 2020;8:278 [14]. C. A representative analysis of sprouting with ImageJ software

6.3.2 In Vitro Sprouting Assay Using Hydrogels of ECM Components

6.3.2.1 Preparation of Hydrogels

Collagen/Fibronectin

1. Stock solution of Collagen type I (Corning, 354,236) is placed on ice for 30 min.
2. The appropriate volume of collagen is transferred carefully into a 1.5 mL eppendorf* in order to have a final concentration of 3 mg/mL (Final volume 3 mg.mL^{-1}/concentration in bottle = volume collagen to be added).
3. DMEM 10X (Sigma, M0275) is added and mixed well with the collagen in order to have a final concentration of 1X (final volume /10 = mL 10X DMEM).
4. 1 N NaOH** must be added gradually, and mixed very well, to achieve a neutral pH (the color of the solution will be pink)***.
5. Fibronectin at 100 ng/mL final concentration is added and mixed well.
6. The following volume of sterile ice-cold dH$_2$O is added: (final volume) – (volume collagen) – (volume 10X DMEM) – (volume 1 N NaOH) – (volume fibronectin) = volume dH$_2$O to add.
7. All the above procedure is performed strictly on ice.
8. 50 μLof the solution are added in a well of a precooled 96 well plate, which is then transferred to the incubator (37 °C) for 30 min, in order for the hydrogel to be formed.

*Collagen stock solution is very viscous; therefore, it needs to be transferred slowly using a precooled tip.

**The volume of NaOH must be calculated based on the concentration of collagen in the stock solution. Collagen concentration can vary between different batches, from 3 to 4 mg/mL. A formula by Corning {(volume collagen to be added) x 0.023 mL = volume 1x NaOH} can be applied in order to calculate the volume of 1 N NaOH required. It is recommended to measure the pH with litmus paper the first time (an acidic or basic environment will have negative impact on the formation of the hydrogel and on the viability of the cells).

***After the pH is neutralized, the collagen can solidify. To avoid that, the vial should be kept on ice all the time and the next steps must be performed quickly.

Fibrin

1. In each well of a 96 well plate 50 μL of the hydrogel will be formed. First the appropriate volume of thrombin will be added to the well*, in order to have a final concentration of 0.64 u/mL.
2. Then the appropriate volume of fibrinogen will be added to the well in order to have a final concentration of 2 mg/mL (Final volume 2 mg.mL^{-1}/concentration in bottle = volume fibrinogen to be added).
3. The following volume of sterile ice-cold dH2O is added: (final volume) – (volume fibrinogen) – (volume thrombin) = volume dH2O to add.

4. It is allowed to clot for 5 min at RT.
5. The 96 well plate is then transferred to the incubator (37 °C) for 20 min, in order for the hydrogel to be formed.

*The final fibrin solution must be prepared in the well, because it polymerizes too fast and it is difficult to transfer a soluble volume from an Eppendorf to the plate.

6.3.2.2 Sprouting Assay

1. 150 μL EGM-2 medium is added to each well containing the hydrogel and a vascular organoid* is transferred using a 200 μL pipette.
2. Medium is changed every 2 days; the sprouting is observed daily and images are taken using a phase-contrast/fluorescence microscope.
3. Quantification is performed using ImageJ software.

*NOTE: More than 1 vascular organoid can be transferred per well, if the interactions between neighboring organoids are studied or the formation of an extended vascular network is desired.

Take-Home Message
We propose a flexible, small-scale 3D organoid-like platform consisting of hPSC-SMC/EC, which is superior to mixed monocells and sole ECs regarding the development of a mature durable vasculature in vitro. These vascular organoids are a defined in vitro model for studying the paracrine interactions between ECs and SMC subtypes that regulate vessel assembly, phenotype modulation, maturation, maintenance, and vessel destabilization in a way that mimics the physiological assembly of the normal vasculature and therefore might serve as a platform for drug development, including estimations of compound preclinical toxicity and potential metabolic liability.

References

1. Heidary Rouchi A, Mahdavi-Mazdeh M. Regenerative medicine in organ and tissue transplantation: shortly and practically achievable? Int J Organ Transplant Med. 2015;6(3):93–8.
2. Carmeliet P, Jain RK. Angiogenesis in cancer and other diseases. Nature. 2000;407 (6801):249–57.
3. Jain RK. Normalization of tumor vasculature: an emerging concept in antiangiogenic therapy. Science. 2005;307(5706):58–62.
4. Nillesen ST, Geutjes PJ, Wismans R, Schalkwijk J, Daamen WF, van Kuppevelt TH. Increased angiogenesis and blood vessel maturation in acellular collagen-heparin scaffolds containing both FGF2 and VEGF. Biomaterials. 2007;28(6):1123–31.
5. Min S, Ko IK, Yoo JJ. State-of-the-art strategies for the vascularization of three-dimensional engineered organs. Vascular Special Int. 2019;35(2):77–89.

6. Schechner JS, Nath AK, Zheng L, Kluger MS, Hughes CC, Sierra-Honigmann MR, et al. In vivo formation of complex microvessels lined by human endothelial cells in an immunodeficient mouse. Proc Natl Acad Sci U S A. 2000;97(16):9191–6.

7. Jain RK, Au P, Tam J, Duda DG, Fukumura D. Engineering vascularized tissue. Nat Biotechnol. 2005;23(7):821–3.

8. Au P, Tam J, Duda DG, Lin PC, Munn LL, Fukumura D, et al. Paradoxical effects of PDGF-BB overexpression in endothelial cells on engineered blood vessels in vivo. Am J Pathol. 2009;175 (1):294–302.

9. Hashizume H, Baluk P, Morikawa S, McLean JW, Thurston G, Roberge S, et al. Openings between defective endothelial cells explain tumor vessel leakiness. Am J Pathol. 2000;156 (4):1363–80.

10. Melero-Martin JM, De Obaldia ME, Kang SY, Khan ZA, Yuan L, Oettgen P, et al. Engineering robust and functional vascular networks in vivo with human adult and cord blood-derived progenitor cells. Circ Res. 2008;103(2):194–202.

11. Derda R, Laromaine A, Mammoto A, Tang SK, Mammoto T, Ingber DE, et al. Paper-supported 3D cell culture for tissue-based bioassays. Proc Natl Acad Sci U S A. 2009;106(44):18457–62.

12. Bhang SH, Cho SW, La WG, Lee TJ, Yang HS, Sun AY, et al. Angiogenesis in ischemic tissue produced by spheroid grafting of human adipose-derived stromal cells. Biomaterials. 2011;32 (11):2734–47.

13. Korff T, Augustin HG. Integration of endothelial cells in multicellular spheroids prevents apoptosis and induces differentiation. J Cell Biol. 1998;143(5):1341–52.

14. Laschke MW, Menger MD. Spheroids as vascularization units: from angiogenesis research to tissue engineering applications. Biotechnol Adv. 2017;35(6):782–91.

15. Markou M, Kouroupis D, Badounas F, Katsouras A, Kyrkou A, Fotsis T, et al. Tissue engineering using vascular organoids from human pluripotent stem cell derived mural cell phenotypes. Front Bioeng Biotechnol. 2020;8:278.

16. Chistiakov DA, Orekhov AN, Bobryshev YV. Vascular smooth muscle cell in atherosclerosis. Acta Physiol. 2015;214(1):33–50.

17. Shepro D, Morel NM. Pericyte physiology. FASEB J. 1993;7(11):1031–8.

18. Regan JN, Majesky MW. Building a vessel wall with notch signaling. Circ Res. 2009;104 (4):419–21.

19. Armulik A, Abramsson A, Betsholtz C. Endothelial/pericyte interactions. Circ Res. 2005;97 (6):512–23.

20. Trkov S, Eng G, Di Liddo R, Parnigotto PP, Vunjak-Novakovic G. Micropatterned three-dimensional hydrogel system to study human endothelial-mesenchymal stem cell interactions. J Tissue Eng Regen Med. 2010;4(3):205–15.

21. Holm A, Heumann T, Augustin HG. Microvascular mural cell organotypic heterogeneity and functional plasticity. Trends Cell Biol. 2018;28(4):302–16.

22. Beamish JA, He P, Kottke-Marchant K, Marchant RE. Molecular regulation of contractile smooth muscle cell phenotype: implications for vascular tissue engineering. Tissue Eng Part B Rev. 2010;16(5):467–91.

23. Hedin U, Thyberg J. Plasma fibronectin promotes modulation of arterial smooth-muscle cells from contractile to synthetic phenotype. Differentiation. 1987;33(3):239–46.

24. Liu LP, Zheng YW. Predicting differentiation potential of human pluripotent stem cells: possibilities and challenges. World J Stem Cells. 2019;11(7):375–82.

25. Takahashi K, Yamanaka S. Induction of pluripotent stem cells from mouse embryonic and adult fibroblast cultures by defined factors. Cell. 2006;126(4):663–76.

26. Kyrkou A, Stellas D, Syrrou M, Klinakis A, Fotsis T, Murphy C. Generation of human induced pluripotent stem cells in defined, feeder-free conditions. Stem Cell Res. 2016;17(2):458–60.

27. Ng ES, Davis R, Stanley EG, Elefanty AG. A protocol describing the use of a recombinant protein-based, animal product-free medium (APEL) for human embryonic stem cell differentiation as spin embryoid bodies. Nat Protoc. 2008;3(5):768–76.
28. Tsolis KC, Bagli E, Kanaki K, Zografou S, Carpentier S, Bei ES, et al. Proteome changes during transition from human embryonic to vascular progenitor cells. J Proteome Res. 2016;15 (6):1995–2007.

Application of Scaffold-Free 3D Models

Sebastian Kreß, Ciarra Almeria, Sabrina Nebel, Daniel Faust, and Cornelia Kasper

Contents

S. Kreß (✉) · C. Almeria · S. Nebel
University of Natural Resources and Life Sciences, Vienna, Austria
e-mail: Sebastian.kress@boku.ac.at; ciarra.almeria@boku.ac.at; sabrina.nebel@boku.ac.at

D. Faust
TissUse GmbH, Berlin, Germany
e-mail: daniel.faust@tissuse.com

C. Kasper
Institute of Cell and Tissue Culture Technologies, University of Natural Resources and Life Sciences, Vienna, Austria
e-mail: cornelia.kasper@boku.ac.at

© Springer Nature Switzerland AG 2021
C. Kasper et al. (eds.), *Basic Concepts on 3D Cell Culture*, Learning Materials in Biosciences,
https://doi.org/10.1007/978-3-030-66749-8_7

> **What You Will Learn in This Chapter**
>
> This chapter will provide an overview on scaffold-free 3D models focusing on aggregates, spheroids, and organoids. While often used synonymously, the differences are presented as well as the origin and advancement until the current date. Moreover, the advantages that scaffold-free systems offer and their limitations are discussed as well as fields of applications ranging from stem cell expansion, pharmacological high-throughput tumor drug screening for personalized medicine, to human-on-a-chip systems facilitating the systemic investigation of drug metabolization reducing animal experiments by considering multiple organ functions within one setup. Furthermore, approaches and requirements for the generation and cultivation strategies are outlined.

7.1   Cells, Assemble!

Scaffold-free 3D cell culture describes the cultivation of cells without the usage of any template structure guiding or forcing anchorage-dependent cells towards a given architecture.

With the advantages of 3D cell culture approaches over 2D cultures illustrated in Chap. 1, some similarities and difference between 3D scaffold-based and scaffold-free are: While for scaffold-based cultures a scaffold made of natural or synthetic biomaterials is given for the cells to adhere, for scaffold-free culture any cell–matrix adherence is prevented to promote cell–cell contact and attachment. Further, for scaffold-based cultures the environment is defined before the cells are seeded onto or into it with a custom-made shape mimicking the tissue they originated from, while for scaffold-free cultures there is no predetermined microenvironment but the cells are brought into such close contact without any other opportunity to adhere to than the neighboring cells initially forming unshaped clusters. Given physiological culture conditions, the cells arrange themselves and produce their own extracellular matrix (ECM) supporting their function by establishing and shaping their microenvironment and providing vital cues by depositing growth factors and binding sites. Yet, in scaffold-based cultures, the cells do also modify and alter the given environment dependent on their needs and function. Further advantages of scaffold-free cultures for anchorage-dependent cells are described below in Sect. 7.3.

Despite single cell suspension culture appearing to be also a scaffold-free 3D cell culture in terms of an available 3D space the cells can float in, there is no possibility for the cells to arrange in orchestration with neighboring cells. The term 3D culture rather refers to the active organization and interaction of anchorage-dependent cells with their environment as well as each other. Within single cell suspension culture, the cells hardly interact with each other by direct physical contact, while within cellular aggregates the cells can form tight bonds enabling the formation of functional structures that go beyond single cell functionalities. Still, 3D aggregate culture can be established from single cell suspension cultures. To achieve 3D cultures without the guidance of a supportive structure, keeping cells adherent close to each other or guiding them towards each other, scaffold-free 3D culture exploits the cellular capability for self-assembly. Furthermore, cells tend to secrete their own ECM proteins in aggregates forming their own microenvironment, which reciprocally affects cellular fate and functionality as well as their further ECM deposition [1]. Moreover, for the formation of aggregates it is also possible to employ a scaffold-based approach seeding a high density of cells into a hydrogel to mechanically support the cellular accumulation and attachment. This is especially often used for the formation of organoids to mechanically support the migration, arrangements, and organization as well as the formation of functional structures until the establishment of a fully functional organoid. However, not all cells possess the capacity for aggregate formation or have lost it during prolonged 2D cultivation.

Generally, the term "aggregate" represents a multicellular entity that is formed via self-assembly of cells with no particular shape. Aggregates can be formed from highly proliferative single cells or from multicellular (co-)cultures. In either case, low adhesion cell culture surfaces promote the self-aggregation of cells into a cluster. While growing, the aggregates condense creating heterogenous gradients of oxygenation, nutrient supply, and hence cellular proliferation and necrosis with increasing size.

In contrast to rather randomly shaped cellular aggregates, spheroids are defined by their spherical shape, which could be formed by the cells themselves or by a controlled or forced manner from aggregates. The generation is easy to establish (explained in more detail in Sect. 7.4), their size is scalable with a high reproducibility.

Lastly, the term "organoid" refers to 3D in vitro tissues that possess multicellular, structural, and functional key properties of an organ at a micro- to millimeter scale. While aggregates and spheroids are formed mainly by cell–cell adhesion, organoids require to undergo further developmental processes driving the establishment of functional structures. Therefore, they often require to be embedded into a hydrogel for structural support while their formation. Organoids are usually composed of cocultures with different cell types to meet the different functionalities of the miniature organs. The different cell types are differentiated from stem or progenitor cells during the culture by exploiting the stem cell capabilities of self-renewal, differentiation plasticity, and self-organization.

Organoids have been used for decades in embryonic stem cell research but have already advanced to organ development and disease modeling using adult or pluripotent stem cells as they demonstrate promising use in medical applications and for investigation of therapeutic interventions due to their in vivo like complexity. There are well-established applications of organoids mimicking function and microanatomy for intestine [2, 3], lung [4], brain [5, 6], liver [7], kidney [8], and retina [9]. Moreover, the diverse organoids can be implemented into microfluidic organs/human-on-a-chip systems. The combination of multiple organotypic tissue-like structures onto one platform enables the investigation of the mode of action of compounds on a systemic level taking the interplay of different organs and their respective metabolic conversion of compounds into account. Such systems are presented in Sect. 7.5.

7.2   Advancement of Scaffold-Free 3D Culture

Organoid culture developed as early as in 1906, when the hanging drop method was established, was only considered as an extension of 3D cell culture. This approach typically referred to the assembly of small tissue fragments including epithelial tissues, which were separated from its stroma by enzymatic digestion or mechanical manipulation, in order to produce different organ-like structures. Organoids research offered the possibility to grow human tissues in a personalized manner, revealing their potential for human biology and medical research. Nowadays, organoids are defined as mini-clusters of cells, formed by self-organization, grown in a 3D environment in vitro and stimulated to differentiate into functional cell types. It allowed to recapitulate morphogenic structure and function of an organ in vivo, obtained from tissue formation generated from any individual by self-organization and spatially restricted lineage commitment and cell sorting. Hereby, various signaling pathways are activated and mediated by intrinsic cues including endogenous factors produced by the cells and surrounding extracellular microenvironment such as the ECM [10, 11].

Various cell types were used for the generation of organoids including embryonic stem cells (ESCs), induced pluripotent stem cells (iPSCs), and neonatal or adult stem cells (ASCs) [12]. At the beginnings of organoid research, the most abundant cell type used was ESC due to their limitless self-renewal in vitro and differentiation capacity to form all three primitive germ layers—mesoderm, ectoderm, and endoderm, as well as germ cells (sperm and ova). Initially, 3D aggregates of ESCs were formed using embryoid body (EB) culture techniques enabling the assembly of complex cell adhesions and intercellular signaling of early embryonic development. Following the aggregation phase, spontaneous formation of a primitive endoderm (PE) layer on the exterior surface can be observed. Reportedly, major cues for the stimulation of PE formation were identified including fibroblast growth factor (FGF) signaling mediated by the PI 3-kinase pathway. Consequently, an epithelial morphology on the EB surface can be distinguished leading to further differentiation into a visceral and parietal endoderm as well as the deposit of a basement membrane, which is rich in laminin and collagen IV. The basement membrane promotes the survival of adjacent cells on the EB surface, whereas the lack of direct contact to it leads to apoptosis of cells and forms a cystic cavity in most EBs. Furthermore, the size of the EB, soluble factors, ECM, and cell–cell interactions all demonstrate to influence ESC commitment. As EB research progressed further, analyses of spontaneous differentiation of EBs from ESC have yielded important insights into the molecules that direct primitive endoderm differentiation and led to the start of expanded organoid cultures of different cell types avoiding the ethical issues that come along with using ESC (see Fig. 7.1) [13].

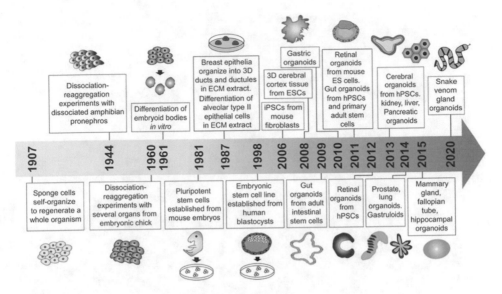

Fig. 7.1 Timeline of organoid culture development. An overview of key landmark breakthroughs leading to the establishment of various organoid technologies. *3D* three-dimensional, *ECM* extracellular matrix, *ESCs* embryonic stem cells, *hPSCs* human pluripotent stem cells, *iPSCs* induced pluripotent stem cells [10]

Due to these milestones made in organoid research, the advancement in biotechnology towards tissue engineering and biomaterials offered the development of many devices, such as microfluidic technologies, in order to support a more physiologic organoid culture with a 3D architecture. Hereby, fluid flow was introduced which enables high-throughput testing, environmental sampling, and biosensing and could create multi-organ models within a microfluidic system. These technologies are continuously exploited for a wide range of tissue types increasing the potential to have a significant impact in medicine with regard to functional differentiation and integrity of form and function maintenance. Furthermore, implementation of organoids could not only be in drug development research but also for patient treatment. Use of organoids has been widely notable for development and disease modeling, precision medicine, toxicology studies, and regenerative medicine (as illustrated in Fig. 7.2).

The most studies were conducted for different types of cancer, where organoids were derived from different murine and human tumors. Tumor organoids have the advantage over 2D cultures that they recapitulate the complexities in cancer tissue. Within the 3D environment of the formed organoid, tumor stem cells can be maintained, as well as their more differentiated progeny mimicking the heterogenicity of the primary tumor cell population which has been shown to be able to predict the clinical drug response. Furthermore, tumor spheroids are easy to generate due to the high proliferative potential of tumor cells facilitating the generation of large amounts of clonally identical aggregates for high-throughput drug screening. Tumor spheroids were first described in the early

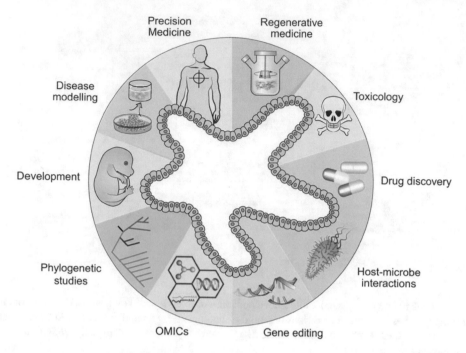

Fig. 7.2 Applications of organoid culture technology. A schematic illustration of various applications of organoids in different areas including precision medicine, regenerative medicine, toxicology, developmental biology, disease modeling, host–microbiome interactions, gene editing, phylogenetic studies, and drug discovery studies [13]

1970s obtained by non-adherent culture of cancer cell lines. The cells are mainly harvested by mechanical dissociation of primary tumor tissue and used for tumor stem cell expansion and as a model for tumor development and drug screening.

As an example, colorectal cancer (CRC) organoids were established from different anatomic sites and indicate to have distinctive sensitivities towards Wnt3a and R-spondin. Furthermore, human liver cancer organoids generated from patient tissues were achieved by extensive refinement of medium conditions to expand the three common subtypes including hepatocellular carcinoma, cholangiocarcinoma, and combined hepato-cellular cholangiocarcinoma. Studies on primary breast cancer organoids have also been abundantly conducted in order to rebuild the parent tumor's morphology, histopathology, and mutational landscape. Numerous organoid models have been established to investigate gastric, prostate, bladder, brain, ovarian, and lung cancers. Nevertheless, the scale and application are limited due to problems with long-term maintenance as well as nutrient and gaseous diffusion. Limitations in media exchange can lead to necrosis as well as lack of in vivo properties such as a defined body axis, a functional immune or nervous system, or a functional vasculature. Although the advances of organoid research have been reported in the past decades towards human-specific aspects of organ development and physiology, it still remains a challenge to uncover aspects of human biology, which highly rely on integrated physiological and complex organ systems [12].

Conversely, a profound number of studies are available at present towards the use of biomaterials (as described in Chap. 2, 4, and 5) in order to provide a 3D microenvironment that mimics that of native tissue. The main requirement is the facilitation of cell–ECM or cell–cell interaction in three dimensions. This is mostly supported by providing a 3D structure, rather than a flat surface coated with motifs facilitating cell attachment and relying on the cells to build up sufficient ECM. It is known that the ECM plays a major role in cellular behavior as it provides a mix of secreted components and defined architecture that stimulates a specific cell response such as cell differentiation. ECM, which is mostly rich in collagen, fibronectin, and laminin proteins, appears to be a potent mediator towards a directed cell fate in organoid as well as in other 3D cell culture models. Furthermore, the incorporation of a high collagen content in EB displayed to bypass the formation of an EB cavity, a high fibronectin content in EB directed the organoid to differentiate into more epithelial and vascular construct. Therefore, it has been increasingly attempted to directly manipulate the composition of the ECM within the EB microenvironment by introduction of individual matrix molecules like collagen and laminin in a soluble suspension during EB formation [13]. Indeed, researchers were successful to generate organoid in laminin-rich hydrogels from single cells of healthy tissues as well as from malignant tumors [14]. However, many biomaterials used for scaffolding are of animal origin, implicating variability and ethical concerns. 3D culture can also be achieved by preventing growth in 2D as well as on a scaffold structure while making use of cellular self-organization. A lot of interest has been directed to a scaffold-free cultivation model using cell aggregates and spheroids to reflect in vivo conditions and bypass certain disadvantages of 3D matrices for cell culture.

7.3   Advantages of Scaffold-Free 3D Culture

The trend of moving from 2D to 3D cell culture models continues since more understanding has been gained in the past decade about the positive influence of physiological cultivation strategies on cells.

The advantage of 2D cell culture, especially for cytocompatibility or pharmacological screening is the high availability of the tested compound when added to the culture medium, as it can directly interact with the cells residing in a flat monolayer. However, the results are often not clinically relevant as in vivo the cells are grouped within a dense cluster of cells and tissue. Furthermore, in 2D culture, the proliferating cells are mostly unintentionally selected towards a cellular population that grows well on the provided surface disregarding the preservation of functional characteristics. Therefore, three-dimensional cultures have shown to mimic the cell's natural environment and hence the results have gained more relevance for clinical applications [15]. Within a 3D environment, co-cultures of cell populations consisting of stem and their more differentiated progeny cells as well as cells with different specialized functionalities can be maintained within one organoid mimicking the heterogenicity of the primary tissue cell population. However, scaffolds can also adsorb or block drugs or compounds hindering them from reaching the cells they are directed towards. In this regard, scaffold-free systems offer a complex 3D structure without a structure, like solid opaque scaffolds, restricting accessibility.

Compared to conventional monolayer cultures, scaffold-free approaches have shown significant improvement with regard to cell viability, growth, and differentiation as well as towards related metabolism, genotype, and phenotype [15]. The cellular behavior in tissues has been suggested to be greatly influenced by diffusive mass transfer due to the propagation of nutrient, metabolic waste products, and oxygen gradients. Mesenchymal stem cells (MSC) have shown enhanced biological function, including differentiation potential, maintenance of stem cell properties, when cultivated as cell aggregates, particularly [16]. It has been observed that MSC experience different strain and rigidity in aggregate cultivation and are therefore required to conform adhesive properties and their phenotype to it, resulting in manifestation of enhanced anti-inflammatory, angiogenic, and tissue regenerative capabilities. With this, cellular behavior observed in such models could seemingly comply to that of a cell's natural behavior, which increases the significance of in vitro models and ultimately the translation into clinics [11, 16, 17]. This culture method is of particular interest for stem cell differentiation and cancer research. Furthermore, aggregates and organoids forming layered and organized structures resemble the complexity of tissues. The following sections outline the generation and cultivation strategies of cell aggregates and spheroids.

7.4   Generation of 3D Cell Aggregates and Spheroids

By utilizing cell-repellent plates, coating surfaces with non-adhesive properties, or hanging drop technique, higher density of cells can be obtained in suspension. Thereby, cellular self-organization facilitates the formation of multicellular aggregates. These 3D spheroid or organoid structures can be formed by various methods. As mentioned in Chap. 1, this is only possible with cells that are able or can be adapted to grow in suspension. However, sole suspension culture is not yet equal to 3D culture. In suspension, the cells do not have a forced two-dimensional growth area but are actually lacking any given growth area.

The generation of aggregates, i.e. spheroids or organoids, is either established by self-organization of the cells or methods supporting the formation of clusters, depicted in Fig. 7.3 and further described below.

Initially, cells-to-spheroid formation kinetics depend on the expression and affinities of cell adhesion molecules such as integrins, cadherins, and ECM proteins, which permit cell–cell contacts and form loose cell aggregates. Next, aggregates enter a delayed reorganization phase and initiate cellular self-assembly. In the last phase, aggregates start compaction due to strong interactions of cadherins between cells, leading to actomyosin-mediated contractility and further to polarization and upregulation of cortical tension at the boundary

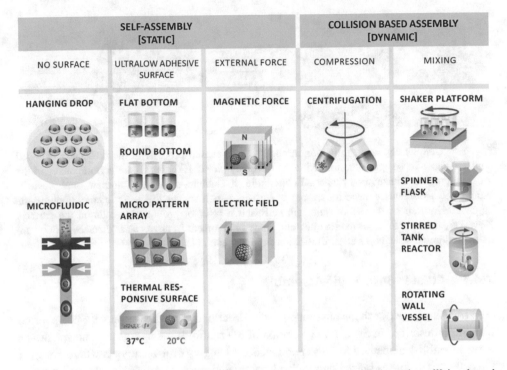

Fig. 7.3 Different approaches for static cluster-based self-assembly and dynamic collision-based assembly of aggregates. Self-assembly of cells can be forced using no or ultra-low adhesive surfaces or external forces. Collision-based assembly is conducted by compression or mixing. Illustration from Egger et al. [16]

of multicellular aggregates. Consequently, this tension dominates over the mechanical energy of cell–cell contacts, resulting in enhanced adhesion among all cells in the aggregate and promote the transition to a compact spheroid (Fig. 7.4). During spheroid formation a necrotic core, represented by quiescent, viable cells, is situated on the surface of the spheroid surrounded by proliferating cells. This phenomenon is developed due to the occurrence of an oxygen, nutrition, and waste product gradient. However, this phenomenon has only been observed in spheroids, with a diameter of >500 µm, cultivated under static conditions. Modulating the dynamics of aggregate formation systematically remains a challenge. Nevertheless, advances in bioengineering strategies demonstrate improved control by modulating cultivation parameters such as rotary speed maintained in a dynamic setting, application of cell surface modifications to enhance cell adhesive properties or cytoskeleton modulators such as cytochalasin D (cytoD), dextran sulfate (DS), and lysophophatidic acid (LPA) to prevent overly compacted aggregates [18, 19].

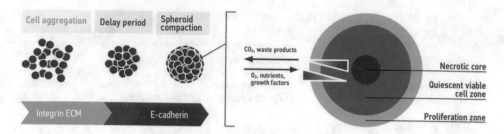

Fig. 7.4 Cellular aggregate formation through a three-phased process. In the first phase cell–cell contacts cause loose aggregate formation. In the delay phase, E-cadherins are accumulated and the loose aggregate initiates compaction. Furthermore, a gradient of oxygen, nutrient, and waste is established from inner core to the aggregate surface, which creates the necrotic core, dividing the aggregate into an inner core of quiescent cells and a layer of proliferative cells on the surface. However, it has been observed that necrotic cores only appear in aggregates (diameter >500 μm) cultivated under static conditions. Illustration from Lin et al. [18]

7.4.1 Cluster-Based Self-Assembly

Hereby, single cell suspensions undergo the described three-phase process of spheroid formation described in Sect. 7.4 Generation of 3D scaffold-free aggregates and spheroids. A well-established method for this approach is the hanging drop technique, where a drop of the cell suspensions is placed on a cell culture-treated surface or dish and subsequently the platform is inverted, allowing the drop to hang from the surface and formation of an aggregate occurs within 24 hours. More sophisticated platforms utilize cell culture plates with cell-repellent properties and different shape of wells (round, flat, micro-patterned) such as round-bottom multi-well plates with ultra-low attachment properties that mediate self-assembly of spheroids (Fig. 7.3). Round-bottom multi-well plates create more uniformly sized and shaped aggregates compared to flat bottom cavities, which makes them more convenient for the generation of comparable aggregates. Based on these observations, various microwell designs composed of different micro-patterned materials including polyethylene glycol (PEG) hydrogels and polydimethylsiloxane (PDMS) are being developed to achieve high yield production of homogenous aggregates in a cost-effective manner. In the following paragraphs, two other types of specialized plate will be described which demonstrates unique features and several advantages over the conventional hanging drop method: *Sphericalplate 5D (SP5D) (Kugelmeiers AG, Switzerland)* is a micro-patterned multi-well plate containing 12×750 microwells ($=9000$ aggregates per plate) (Fig. 7.5). A single cell suspension is dispersed in each well treated with high-end ultra-low attachment nanocoating which allows formation of uniform aggregates within 24 h. The generated aggregate size can be precisely controlled by simply adapting the seeding density and maintenance of cell aggregates during cultivation is easily performed. Subsequently, a higher standardization and convenient upscaling is enabled while assuring labor-efficient

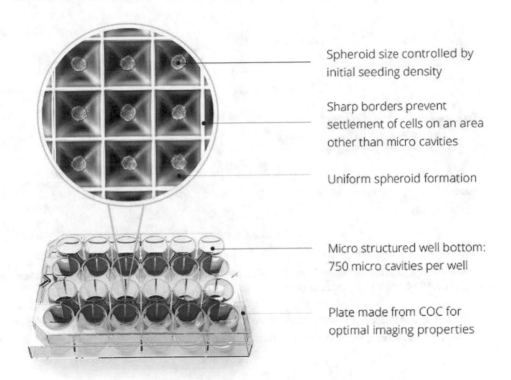

Spheroid size controlled by initial seeding density

Sharp borders prevent settlement of cells on an area other than micro cavities

Uniform spheroid formation

Micro structured well bottom: 750 micro cavities per well

Plate made from COC for optimal imaging properties

Fig. 7.5 Sphericalplate 5D (SP5D) for generation of large quantities of spheroids. The SP5D has micro-patterned microwells with non-adhesive surface coatings to allow uniform and size-controlled cell aggregation of up to 9000 spheroids/plate

and cost-effective advantages. This platform facilitates the translation from bench to bedside and ensures reliable as well as safe application of cell therapy and diagnostics.

GravityPlus™ (InSphero, Switzerland) is a special hanging drop cell culture system that facilitates self-aggregation without artificial matrices. Single cells undergo the conventional three-phase aggregation process via the hanging drop cultivation technique as depicted in Fig. 7.6. Briefly, cell suspensions are introduced into the plate's unique SureDrop™ inlets (of individual wells using a manual single pipette or a multi-channel pipette system) in order to direct a formation of hanging drops at the outlets under the plate by gravity and the pressure caused by flow of liquid. Through this, same sized aggregates are produced in a controllable manner within 2–4 days. However, these hanging drops and thus the aggregates residing within them can easily be destroyed by careless handling of the plates, which detach the droplets from the wells. Therefore, after the desired aggregate size or compaction has been reached, the droplet can be transferred into the special non-adhesively coated 96-well microtiter plate (GravityTrap™). This platform reduces the risk of losing the spheroids during handling like medium changes. Thus, it allows for a safe long-term cultivation of aggregates and facilitates downstream processes.

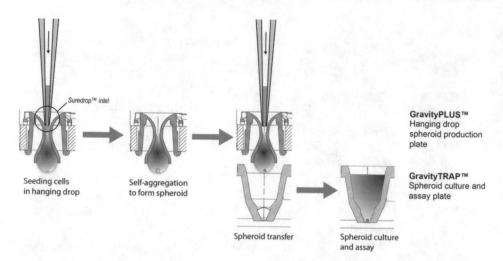

Fig. 7.6 Cell aggregeate formation in the GravityPlus™ plate and transfer into the low-attachement GravityTrap™ plate for long-term cultivation and analysis

Furthermore, automated and microfluid-based high-throughput hanging drop cultivations are aimed to be developed to obtain reproducible aggregates in time-efficient manner.

7.4.2 Collision-Based Assembly

Traditional techniques for spheroid formation such as spontaneous aggregation via the hanging drop method are very labor intensive, do not provide large quantities of aggregates, indicate serial passaging challenges, and offer limited control over size and shape of aggregates. Alternative approaches have been developed to address these matters by introducing collision-based or forced aggregation, wherein shape and size of single aggregates can be directly controlled. Strategies for dynamic forced cell aggregations are depicted in Fig. 7.3. The centrifugation method is performed by transferring single cell suspensions into microwell arrays and subject them to centrifugal forces for the generation of size-controlled aggregates. However, strong centrifugation force can lead to undesired and spontaneous effects such as altered cellular viability and differentiation as well as stem cell properties in case of MSCs. This strategy has been widely used for chondrogenic differentiation, also referred to as pellet or micromass culture, which enhances the potential of MSC to differentiate into chondrocytes. Other techniques rely on mixing of the cell suspension in spinner flasks, stirred tank bioreactors, and rotating wall vessel, which resulted in less uniformly sized aggregates compared to aggregates formed in ultra-adhesive multi-well plates placed on a shaker platform. A combination of such multi-well plates and subsequent centrifugation resulted in yielded aggregates with controlled size and homogenous shape distribution. Over a decade, researches have intended to further optimize these strategies to obtain highest yield and homogenously formed aggregates without compromising the cells' attributes [18].

7.5 Cultivation Strategies of Aggregates

Several studies have reported increased viability, proliferation, and differentiation potential as well as paracrine activity and effects of cells cultivated in a dynamic setting such as fluid flow and stirring. Dynamic cultivation systems such as stirred tank bioreactors, perfusion bioreactors, horizontal or orbital shaking platforms demonstrate to have substantial positive impact on the cellular behavior compared to conventional static cultivation. Furthermore, studies suggest that the problematic necrotic inner core could be eliminated by converting from static to dynamic cultivation. Specifically, MSC aggregates cultivated in a stirred tank reactor have demonstrated to maintain stemness and an undifferentiated state even after 16 days of dynamic cultivation. Downstream process analyses resulted in robust trilineage differentiation, no alterations of MSC markers as well as the absence of a necrotic core [20]. Wherein, it was suggested that the formation of cell aggregates could have shielded shear forces to reach the inner core and consequently avoided spontaneous differentiation of aggregates during a long-term cultivation subjected to mechanical stimuli. A schematic summary of the effects of static and dynamic conditions is illustrated in Fig. 7.7. Nevertheless, it is still not commonly applied for cell aggregate cultivation as only a few studies conducted dynamic conditions for aggregate culture. Furthermore, no definite statements have yet been made regarding an increase of the differentiation potential of 3D dynamic cultivation in comparison to 3D static cultivation. As of today, comparisons between 3D dynamic cultivated cells and 2D static cultivation towards gene expression, secreted proteins, cytokine levels, and survival rate after implantation have been investigated predominantly. Therefore, future research might focus more on a direct comparison between static and dynamic cultivation of cell aggregates and spheroids to evaluate its therapeutic potential and may verify this approach as an alternative to monolayer cell culture techniques [16, 21].

Facilitating dynamic cultivation, bioreactors mimic physiological culture conditions, improve nutrient supply, especially important the larger the aggregates and the higher their cell density, as well as apply mechanical stimuli. The operating principle of common bioreactors applicable for scaffold-free 3D cell culture, particularly mixing bioreactors, is described in more detail in Chap. 2. Here, we want to focus on actual applications in different research questions, combination of cultivation strategies, and the parameters important for expanding, maintaining, or differentiating 3D cell constructs. Rather than attempting to cover the whole field, a hand full of different research projects are presented in more detail for illustration. A mixture of various cultivation techniques, tissue types, and

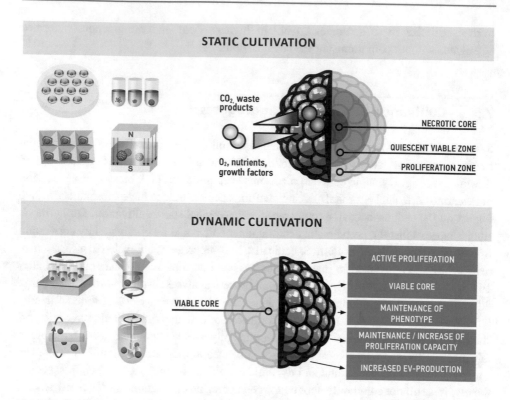

Fig. 7.7 Comparison of effects observed in static and dynamic cultivation of MSC aggregates. The cell aggregates have been suggested to create three layers due to the development of oxygen, nutrient, and waste product gradients under static conditions: a necrotic core in the center, a quiescent viable zone of non-proliferative cells, and an outer layer with proliferating cells. On the contrary, dynamic cultivation conditions have been reported to maintain a viable core with actively proliferating cells throughout the cultivation. Furthermore, cells from the aggregates cultured under these dynamic conditions reveal to maintain their phenotype, proliferation, and differentiation capacity [16]

intended applications was chosen to emphasize the wide range of possible applications of 3D cell culture. A quick overview of the discussed projects can be found in Table 7.1.

7.5.1 Static Cultivation

Despite the above discussed benefits of dynamic cultivation some setups still use static conditions due to handling efficiency or possible negative effects on the cell constructs. Especially in the first instances of aggregation, as described in Chap. 1. 4. (Cluster-based self-assembly), the cells have to be kept under static conditions in many cases until they form enough cell-to-cell contacts to become stable enough and maintain their shape in further handling steps. In a 2011 paper by Hildebrandt et al. [41], different MSC spheroid

Table 7.1 Overview of different 3D cultivation strategies

Tissue/final graft	Intended application	Cells	Organism	Aggregation technique	Cell number	Microtissue/graft size	Cultivation duration	Cultivation	Ref
Ex vivo skin/in vitro skin model and hair coculture	Cosmetic + drug testing	[EpiDermFT™] + juvenile prepuce biopsy + follicular unit extracts (FUE)	Human	N/A	N/A	Surface area of trans-well 14.3 mm2	7–14 days	Perfusion (microfluidic chip)	[22]
Cell aggregates/single cells	Cells for cell based therapies (CBT)	Human MSC (hMSC)	Human	Collision-based assembly	10^5/mL	N/A	6 days	Stirred tank reactor	[23]
Cell aggregates/single cells	Cells for cell based therapies (CBT)	Human iPSC(hiPSC)	Human	Collision-based assembly	$0.33–2 \times 10^5$ /mL	Aggregate Ø 140 μm	4–5 days/passage	Dynamic mixing (BioLevitator™ system)	[24]
Skin (commercially available)	Cosmetic + drug testing	Keratinocytes + fibroblasts	N/A	Trans-well culture	N/A	N/A	N/A	Static	[25]
Small diameter tissue-engineered blood vessel	Drug screening and toxicity studies	hMSC + human endothelial progenitor cell (hEPC)	Human	Cell sheet formation with subsequent layering (wrapping) on support mandrel	N/A	Tube of 30 mm length; outer Ø 1.13 mm; inner Ø 1 mm	21 days static +14 days dynamic (on mandrel) + 7 days dynamic (in reactor)	Static + rotating wall reactor + perfusion	[26]
Myocardial microtissue	Cardiac tissue engineering	Primary cardiomyccytes + primary myocardial fibroblasts	Mouse/rat	Hanging drop	2.1×10^3 / microtissue	130–320 μm	6 days	Static	[27]
Vascularized myocardial macrotissues	Cardiomyocyte-based cell therapy for myocardial infarction	Neonatal (NRCs) – adult (ARCs) rat cardiomyocytes	Rat	Hanging drop + microtissue fusion	2.5×10^3- 1×10^4 / microtissue	180–400 μm (microtissue) cylinders with 3 mm Ø (macrotissue)	7 days hanging drop +7 days static	Static	[28]
Small diameter tissue-engineered blood vessel	Vascular graft for coronary bypass	Human artery-derived fibroblast (HAFs) + human umbilical cord-derived endothelial cells (HUVECs)	Human	Hanging drop + microtissue fusion	$10–11 \times 10^3$ / microtissue; $4–5 \times 10^3$ microtissues per graft	Tubes of 5 mm length; 1 mm wall thickness	7 days hanging drop +14 days in bioreactor	Pulsatile flow	[29]

(continued)

Table 7.1 (continued)

Tissue/final graft	Intended application	Cells	Organism	Aggregation technique	Cell number	Microtissue/graft size	Cultivation duration	Cultivation	Ref
Cardiomyocytes	Bioartificial cardiac tissue formation	HES3 NKX2-5$^{eGFP/w}$ + hCBiPS2 + hHSC_F1285T_iPS2	Human	Collision-based assembly	3.3×10^5/ml	531.6 μm	4 days preculture +10 days differentiation	Dynamic mixing (DASbox system)	[30]
Adipose microtissue	Drug testing + metabolic research	Primary adipocytes	Human	Hanging drop	5-20×10^3	N/A	49 days	Static	[31]
Cortical neurospheres	Basic neurobiology research + drug testing	iPSC line StemUse101 (derived from peripheral mononuclear cells (PBMCs))	Human	Collision-based assembly	2.5×10^5/mL	230 μm	32 days	Dynamic mixing (DASbox system)	[32]
Cerebral organoid	Basic neurobiology research + disease models	H9 ESCs (ES) + iPS cells	Human	Embryoid body (EB) formation in ultra-low attachment plates	4.5×10^3/EB	N/A	6 days embryoid body formation, 5 days low adhesion plates, 4 days static embedded in matrigel 30 days in spinning bioreactor	Static + spinning bioreactor	[33]
MSC "bullets"	CBT for myocardial infarction	Human umbilical cord mesenchymal stem cells (ucMSC)	Human/rat	Self-assembly in non-adherent culture dish	N/A	40-100 μm	1 day	Static	[34]
Neocartilage graft	Biological replacements	Primary articular chondrocytes	Human/bovine	Self-assembly in custom agarose mold	7-8×10^6	13 mm × 8 mm	5 days	Tensile strain	[35]
Liver-kidney multi-organ-on-a-chip	Drug testing (hepato- and renal toxicity)	HepaRG cell line + proximal tubule cell line RPTEC/TERT1 + HHSteC cell line	Human	Trans-well culture + dynamic spheroid aggregation (spheroid plates on shaker)	N/A	Chip: 76 × 25 × 3 mm	16 days	Dynamic aggregation on shaker (liver spheroids), trans-well culture (renal barrier) followed by microfluidic slide	[36]
Hair follicle	Hair regeneration + drug testing (chemotherapy)	Dermal papilla (DP) cells + hair follicle melanocytes + keratinocytes	Human	Self-organization in ultra-low attachment plates	25,000/cm2	150 μm (initially)	7 days (DP cells only) + N/A (triculture)	Static	[37]

4-organ-chip (skin, liver, kidney, gut)	Drug testing	Human small intestinal barrier models [EpiIntestinal™] + HepaRG cell line + human primary hepatic stellate cells (HHStecC) + juvenile prepuce biopsy + hTERT immortalized human renal proximal tubular epithelial RPTEC/TERT1	Human	Trans-well culture + dynamic spheroid aggregation (spheroid plates on shaker)	N/A	Chip: 76 × 25 × 3 mm	28 days	Pulsatile perfusion (microfluidic chip)	[38]
Blood–brain barrier	Basic research on 3D confirmation, future: Model for drug delivery	Human primary brain endothelial cells (hpBECs) + primary pericytes (hpPs) + primary astrocytes (hpAs)	Human	Hanging drop coculture	$1\text{-}3 \times 10^3$	N/A	5 days	Static	[39]
Artificial liver lobule	Pharmacodynamics/ personalized medicine	Primary hepatocytes + primary hepatic stellate cells	Rat	Guided seeding onto PDMS membrane	2.5×10^5	N/A	14 days	Perfusion (microfluidic chip)	[40]

formation techniques were compared to each other, where aggregation in 96-well-plates and further cultivation in same proved to be the easiest handling-wise and the most consistent technique with regard to spheroid sizes. In comparison, spheroid formation in non-treated bacterial dishes resulted in heterogeneous aggregate size and shape. The same goes for long-term culture of hanging drop spheroids, which requires transfer from the drops to a suspension culture, due to media consumption. Within the suspension, although they had already compacted to size-defined spheroids previously, they still fused with each other and formed large irregular aggregates. An option to circumvent spheroid fusion or reattachment to the plate after initial aggregation (especially MSC are known to notoriously attach to many surfaces labeled as non-adherent) is to embed them into hydrogels or use semi-solid (highly viscous) culture medium [41, 42].

Opting for static conditions could also be ascribed to no need for long cultivation times. For example, MSC spheroids intended for the treatment of myocardial infarction (MI) were implanted into rat MI models already 24 h after aggregation in a non-adherent culture dish [34]. Within shorter timeframes and smaller aggregates, the risk of hypoxic conditions in the core is lowered, considered as one of the major restrictions in cell aggregate cultivation. In the case of chemotherapeutic drug screening however, it actually can be an advantage. It is well known that such a pathophysiological hypoxia better resembles the in vitro patterns within tumors and that it influences drug resistance. Additionally, the expression patterns of hypoxia inducible factor 1 alpha (HIF-1 α) regulated genes are altered indirectly in the cells [43]. Another aspect is that dynamic culture conditions require large quantities of media which conflicts with the limited amount of the tested novel drug.

One example where static 3D cell culture brings novel insights for research compared to the 2D counterpart are spheroid cultures of white and brown adipose tissue. In recent years, it became clear that the purpose of adipose tissue lies not exclusively in the storage of energy in the form of triglycerides but has a large role in maintaining a normal metabolism. Disfunction in this tissue type is related to the development of diabetes type II or cardiovascular disease. So far 2D cultures could not successfully mimic adipose tissue maturation nor final function. Especially primary brown adipocytes lose their differential expression profile rapidly in monolayer culture. However, using hanging drops to generate spheroids from 5000 to 20,000 cells of primary adipose tissue has shown to increase their adipogenic secretion profile. Further, differentiation of pre-adipocytes to mature cells within this 3D setup produced unilocular lipid droplets as seen in vivo compared to multilocular lipid droplet formation in 2D cultures. Combined, these results propose an easy protocol for generation of microtissues that can be used for studying diseases and potential drugs. Prolonged culture of 30 days and potentially more can even be used for screening of chronic exposure [31]. Not only the numbers of diabetes type II cases are at an all-time high, cardiovascular disease is the number one cause of death globally. Heart-related insufficiencies are especially difficult to treat as the regeneration capacity of cardiac tissue is limited. Testing in rodents and establishment of in vitro cardiac tissue are expected to pave the way for future treatment strategies. In this sense tissue function like synchronized beating and also integration of vascularization is a prerequisite for both in vitro models and potential tissue replacements. Generation of microtissues was accomplished by hanging drop culture of neonatal rat (NRCs) and mouse (NMCs) cardiomyocytes, which started to contract after 4 days of cultivation. They kept their tissue specific key characteristics, a sophisticated ECM, high degree of self-organization, and produced VEGF in a size dependent manner [27]. An important aspect considering that this can mediate connection to the host capillary system or induce neovascularization in coculture microtissues. This led Kelm et al. to the implication that such microtissue (μm^3 scale) approaches have the potential to be scaled up to macrotissue (mm^3 scale), a crucial step towards large artificial tissues and therefore clinical relevance. In a further step, they extended their protocol by coating the cardiomyocyte spheroids with human umbilical vein endothelial cells (HUVECs) and let 300 of the resulting microtissues fuse together inside a cylindrical agarose mold to create a macrotissue. They could still observe coordinated contractions and following transplantation of the myocardial patches onto chicken embryos connection to the host vasculature. Further, implanted patches were able to integrate into the rat's myocardium [28]. Staying on the topic of vascular engineering, one desperately sought-after in vitro model is that of the cerebral microvasculature or more commonly referred to as the blood–brain barrier. Highly selective molecular transport into and out of the central nervous system is made possible by the complex architecture of endothelial cells, pericytes, and astrocytes. Brain vascular pericytes (hpPs) are embedded in the abluminal basement membrane, brain endothelial cells (hpBECs) on the luminal side of the vasculature, and cerebral astrocytes (hpAs) in contact with all other types via their endfeet. In order to test drug diffusion through this barrier trans-well culture setups are used. These models are somewhere in between 2D and 3D culture, scaffold-free and matrix-based, as cells are seeded in monolayers on the top (hpBECs) and bottom (hpPs) of the collagen coated trans-well insert, with astrocytes residing in the lower culture compartment. However, Urich and colleagues [39] compared this culture to both Matrigel embedded cells and hanging drop coculture of these cells. What they could observe was that the cells, cultured in a hanging droplet culture plate allowing for more stable droplets and addition of culture medium to the drop, spontaneously self-assembled to organoids, with astrocytes in the core, endothelial sides on the outside, and pericytes as an intermediate layer. These results demonstrated that cell-to-cell communication is enough for functional assembly. Further adaption of this protocol is required for studying functional aspects like transport mechanisms across the blood–brain barrier as is possible in trans-well format.

An already widely accepted cultivation strategy using trans-wells is the formation of partial or full thickness skin grafts. So far, they have been commercially available since the early 2000. EpiSkin™ (L'Oréal), EpiDerm™ (MatTek), and Phenion®FT (Henkel) are some examples [25]. Later two are considered full thickness as they combine keratinocytes and fibroblasts to mimic both epidermis and dermis. Limitations of these models include the incompletion of the complex composition of the dermis in vivo: vascularization, innervation, hair follicles, lymphatics, and sweat glands. Moreover, the barrier function is way lower than in the physiological system, making uptake studies unreliable [25].

At least for one of the shortcomings, a solution may be found. In a 2011 short communication Lindner et al. describe the formation of hair follicle organoids that are even able to produce fibers. Organoid is the term given to the smallest functional unit of an organ, like a sinusoid in the liver. Hair follicle organoids are not of scientific interest to treat pattern baldness or alopecia, especially chemotherapeutic drug induced ones. They observed dermal papilla (DP) cells condensate in ultra-low attachment plates to ~150 μm aggregates. These then initiated ECM production, which could be accelerated by addition of ECM proteins (collagen IV, fibronectin, and laminin), leading to a dermal papillae phenotype. As hair production is mediated by keratinocytes, they in combination with hair follicle melanocytes were added to the organoids. Following the development of tissue polarity and concentric layers, hair like fibers were produced by around 13.5% of neopapillae. Although the composition of the fiber was not determined within this article, it very much resembled human unpigmented hair [37].

7.5.2 Dynamic Cultivation

Still, static culture has its limitations, especially when it comes to expansion of cells, scalability, and GMP compliance. iPSCs and somatic MSCs are promising candidates for a wide range of future disease treatments or modeling systems. However, their prolonged expansion in 2D setups has shown to decrease their anti-inflammatory and differentiation capacities, which led the scientific community to search for alternative cultivation strategies that are cost-efficient, simple, and robust. Multiple groups have reported successful expansion of cells in mixing reactors, foremost stirred tank reactors. Elanzew et al. [24] tested their protocol for prolonged expansion over ten passages of hiPSCs, based on collision-based assembly during dynamic culture in the BioLevitator (see Chap. 2), the group around Abbasalizedah [44] cultivated their iPSCs successfully over 20 passages. The majority of such stem cell expansion strategies use stirred tank bioreactors that initially were developed for bacterial culture and CHO cell culture. Yet following aspects have to be considered to allow for expansion of the more demanding iPSC or ESC stem cell aggregates: "(1) enzymatic dissociation of aggregates and passaging as single cells using different enzymatic dissociation mediums, (2) cell inoculation (3) hydrodynamic culture conditions with different agitation rates and medium viscosities, (4) oxygen concentration, and (5) aggregation kinetics under dynamic conditions" [44]. In order to overcome the natural size limit of aggregates they are disassociated to single cells between passages. Hereby, usually Rho-associated, coiled-coil containing protein kinase (ROCK) inhibitors, for example, Y-27632 are added, to prevent dissociation-induced apoptosis (anoikis) and hence improving cell survival. This is also improved by optimizing inoculation cell numbers, which range usually from $0.03\text{-}1 \times 10^6$ cells/mL, and time-displaced stirring start or lower starting speeds. Notably, all trials mentioned in this study [44] used chemically defined media omitting the addition of FCS, which is a prerequisite for future clinical applications. Using a custom-built continuously stirred tank reactor comprising of a round-bottom glass vessel and a PEEK stirrer, Egger et al. [23] tested scaffold-free cultivation of MSCs. They can be found in high abundance still in adult tissue and usage is not accompanied by ethical concerns as is for ESCs. It is well known that they are capable of multilineage differentiation and possess immunomodulatory capacities. Recent studies showed that hypoxic conditions as well as aggregate cultivation can enhance these characteristics. The study aims at proving that hMSC aggregate culture in stirred tanks is feasible and can be further improved by applying physiological oxygen levels. They achieved a 1.8- and 2.2-fold increase of cells after 6 days of cultivation with 600 rpm under normoxic and hypoxic conditions, respectively. The rather high agitation speed was necessary to prevent cells sinking to the vessel bottom or attachment to the vessel wall. A single cell solution was used for seeding and spontaneous aggregate formation was observed around day 3 of cultivation. Analysis of the dissociated cells after cultivation confirmed that they still exhibited stem cell surface marker and kept their lineage differentiation potential. In summary, these three-dimensional and dynamic cultivation strategies lay the foundation for further advances in regenerative medicine and tissue engineering by providing a scalable and GMP compliant system for producing the building blocks of potential treatments—stem cells.

Mixing bioreactors are not only employed solely for expansion. Kempf et al. [30], for example, have combined an initial expansion step of hPSC with subsequent initiation of cardiomyocyte differentiation protocol. They could successfully show that within the same vessel initial proliferation and aggregate formation and further differentiation to contracting cardiomyocytes are feasible. Another application of a stirred vessel to maintain a 3D cell construct is from 2014, where Lancaster et al. were able to generate cerebral organoids [33]. They used embryoid bodies (EBs) to generate neuroectodermal tissue. These EBs, which were embedded into Matrigel and cultivated in a stirred tank reactor, were still viable after 10 months but could potentially survive indefinitely. The cerebral organoids increased in size until 2 months to around 4 mm in diameter, with complex heterogenous architecture and a fluid filled pseudo-ventricle. In a 2018 paper, expansion and subsequent neuronal differentiation could be combined [32]. HiPSCs were cultured in a DASbox Mini Bioreactor System and after a well-defined protocol of continuous media exchanges and step-wise exchanges of induction and adaption media lasting 22–32 days 2×10^8 neuronal stem and progenitor cells could be harvested—quantities that prove such technologies are able to produce sufficient amounts necessary for applications like drug testing. Upscaling which is feasible could even generate enough material for patient treatments.

Next to mixing, perfusion reactors also have great potential for usage in 3D cell culture. Tissue-engineered vessels are covered as example in this chapter not only because they are essential for every larger 3D cell culture undertaking that shall survive in vitro in the future, but because of their obvious connection to perfusion. Cardiovascular diseases are a big problem in our aging society and with increasing numbers of surgical interventions in coronary heart disease also the number of suitable vessel grafts that are needed in bypass surgery is growing. Already in 1998, the group around L'Heureux was able to demonstrate that small vessels can be engineered without the use of any scaffolding material [45], thereby reducing the risk of detrimental immunological responses. In 2006 he could validate his vessel in a primate model, however the time it took to produce the graft was more than 4 months [46]. By combining multiple 3D cell culture strategies, another team of researchers tried to speed up and improve the generation of a so-called living small diameter tissue-engineered blood vessel (TEBV). As was mentioned earlier, spheroids or microtissues can be used to create larger (mm^3 scale) cell grafts that may become relevant for clinics. A combination of human artery-derived fibroblasts (HAFs) and human umbilical vein endothelial cells (HUVECs) was aggregated using the hanging drop method. After the initial static culture 4000–5000 spheroids were placed directly into the custom-built perfusion reactor to fuse in a tubular shape. This was achieved by the geometry of the device which comprised of an inner silicone tube, 1 mm spacer rings on top and bottom, and an outer porous casket, where the microtissues were seeded within. The inner tubing is perfusable with a pulsatile pump creating mechanical stimulation on the inner side of the vessel graft, this stimulus was applied for 14 days. The resulting TEBV, 3 mm in diameter with a wall thickness of 1 mm, exhibited a layered structure similar to the tunica media and adventitia and enhanced ECM production in only 3 weeks of cultivation [29]. Following L'Heureux concept, TEBVs produced by stacking of cell sheets on top of each other were still a topic in a 2015 publication. Herein, human MSCs (hMSCs) were cultured on nanostructured PDMS molds for cellular alignment; once they had created a stable cell sheet wrapped around a 1 mm thick mandrel in 4 layers and cultured for 2 weeks in a rotating wall bioreactor. After this maturation step the mandrel was removed and human endothelial precursors (hEPCs) were seeded into the created lumen. The graft was mounted in a perfusion bioreactor and continuous flow was applied for a week. Including the 3 weeks prior needed for the cells to become a stable cell sheet in the first step, again the procedure takes more than 6 weeks; not even included are the cell expansion prior to the graft fabrication. Not surprisingly, the authors propose their protocol for establishing in vitro drug screening models rather than actual tissue replacement. To prove this concept, they not only show the maturation of the vessel architecture and MSCs differentiation towards a vascular phenotype but also biological functionality of their construct. Vasodilation and vasoconstriction were assessed by increase of flow rate and addition of phenylephrine, a vasoconstrictor, respectively. The created vessel responded in a similar fashion as native porcine femoral vein, affirming the potential applicability in pharmaceutical studies [26].

In the context of perfusion, we also have to include microfluidics. Although it is usually associated with very low cell numbers and typically cell monolayers it has become a versatile cultivation strategy for scaffold-free 3D spheroids and organoids. For further reading on microfluidic in general (basic concept, fabrication techniques, and equipment) go to Chap. 8. Application of different perfusion regimes at this scale can even facilitate tissue organization. For example, Weng et al. [40] seeded hepatocytes and hepatic stellate cells (HSCs) into a hexagonal cultivation chamber. The chamber has inlets at each corner of the hexagon and an outlet directly in the center above the cells. This setup mimics the flow from the portal vein to the central vein. They observed that after 7 days an organotypic structure with a sinusoid wall-like morphology had formed that represents hepatic physiology. The results show great potential for studies on drug induced liver injury, an important aspect of drug development. Most pharmaceuticals are metabolized and cleared from the organism by the liver, making the need for elaborated functional 3D tissue equivalents obvious.

Perfusion can also improve the so-called ex vivo cultures. As we have discussed earlier, although commercially available, tissue-engineered skin has its limitations. Therefore, skin biopsies as test models are still in use, they are supplied with culture medium, kept in air–liquid interface, and have the advantage of containing all functional subunits next to their barrier functions needed to test penetration of topically applied substances. However, for ex vivo models the cultivation time is rather short. Ataç et al. could show that there is a distinct difference in the integrity of the tissue graft in static versus perfused cultures. In his microfluidic chip a compartment compatible with a trans-well inlay was used to culture the skin biopsies in an air–liquid interface. He also demonstrated that commercially skin grafts can benefit from dynamic cultivation [22].

Size limitations due to the lack of vascularization have been mentioned before (more details in Chap. 6). In some of the previously mentioned culture strategies the application of dynamic cultivation techniques improved the viability and increased the time the graft could be kept in culture. Now, within microfluidic systems we can resolve on top of that an aspect even more important than just nutrient supply. In vitro drug tests, even if applied to organoid tissues, cannot give a complete picture on potential systemic effects because there is no circulatory system that connects these isolated model organs with the rest of the organism. With the use of microfluidics and a vast array of different tissue model cell grafts application relevant system effects can be tested. Preclinical drug studies in animals are systemic but not human, but in vitro testing is indeed human but cannot produce any systemic effects. Here, microfluidic multi-organ-chips (MOCs) can come in handy. Two organs that play a major role in system drug toxicity are liver and kidney, hepatic tissue is responsible for metabolizing xenobiotics, while nephrotic tissue is responsible for the removal from the body. As the liver is responsible for toxification or detoxification processes, it alters the concentration of the xenobiotics, thus greatly influencing toxicity on kidney tissue. In order to simulate this delicate interplay of the two organs, human liver spheroids and a renal proximal tubule barrier were combined on a MOC. Hepatic lobule equivalents were made by co-culture of the hepatic cell line HepaRG with hepatic stellate (HHSteC) cells in 384-well spheroid plates. In this culture protocol the plate is then placed on a shaker for 3 days inside an incubator to produce compact round spheroids. Renal barrier model was crafted by addition of hTERT immortalized human renal proximal tubular epithelial (RPTEC/TERT1) cells to a trans-well insert. These two tissue equivalents were then combined on a 2-organ chip with no direct contact but an interconnected media circuit and kept in culture for 16 days. While the organoids could be kept in a steady state over the culture period, static controls exhibited tissue disintegration of liver spheroids and renal tubule barrier. To further show the ability to perform toxicity studies in this setup, administration of cyclosporin A and immunosuppressant with toxic effects on both liver and kidney was performed on 14 days of culture. This drug was chosen due to high variance between toxicity in different species. The group could show that the organoids expressed tissue specific markers and that their model could simulate drug metabolism and toxicity with the possibility to discriminate toxicity between the two model organs [36]. An even more complete system was presented by addition of skin and gut tissue on to create a 4-organ-chip. With this an ADME (adsorption, distribution, metabolism, and excretion) model for future drug development was realized. Liver and kidney equivalents were made as previously described, for skin integration human juvenile prepuce was used and lastly a commercially available reconstructed human intestinal barrier (EpiIntestinal™) was used as a gut surrogate. The different tissue types were placed in respective compartments on a microfluidic chip, equipped with two separate microphysiological fluid flow circuits overlapping in the renal barrier compartment mimicking a surrogate blood circuit and an excretory circuit. The architecture of the chip allowed for submerged cultures for the intestine and culture of the skin at the air–liquid interface. Three different media were used to establish gradients across the system: small intestine medium on the upper lumen of the gut model, liver tissue medium in the "blood circuit," and proximal tubule medium in the "excretory circuit." The gut model was placed above the blood circuit, creating a barrier for mimicking xenobiotic absorption into the blood. By an incorporated micropump these can be transferred to the liver surrogate, where they would be metabolized. Further on, the renal tubule barrier model separates blood and excretory circuit. The skin model was added to either test for toxicity or as an alternative absorption option. Tissue architecture, cell behavior, and protein expression were then monitored over a cultivation time of 28 days to test the stability and functionality of this four-organ-microchip. Remarkably, glucose levels remained very stable within the three different media indicating barrier functionality; the "blood sugar levels" even resembled in vivo levels. Immunohistochemical analysis also revealed distinct physiological architecture of all four tissue types, including 3D villi like structures in intestinal culture, stratified stratum corneum in skin, and functional polarized membrane in renal tube cultures. Tissue homeostasis between all four tissue types was also verified by gene expression analysis, proofing their applicability in ADME studies [38].

Such MOCs represent a promising alternative to animal testing and assuming that we are able to integrate tissue micro-equivalents of all relevant organs of the human body onto a microfluidic chip, so to say "Humans-on-a-Chip" might even replace them. This is exactly the vision of the 2010 found company "TissUse." They offer previously described 2-,3-, or 4-organ chips with a wide range of different tissues and combinations, customizable to the intended application. As a future perspective the technology might eventually be adapted to screen for personalized drug efficiency by using patient derived cells. Even more excitingly, Humimic Chip XX and XY which will reproduce a systemic model of a complete female or male body are currently in development [47].

A different approach to dynamic cultivation parameters, next to improved cell viability, is the improvement of differentiation. By using dynamic conditions, mechanical stimulation mimicking the in vivo environment can be applied to the tissue constructs.

Cartilage is of great interest to tissue engineers, due to the high clinical relevance of osteoarthritis and other age-related skeletal disorders. Primary chondrocytes are known to lose their differentiated phenotype rapidly in monolayer culture; therefore, 3D culture is widely used in this field already. Because of the low cell to ECM ratio in vivo, tissue-engineered cartilage often takes advantage of various scaffold materials where the cells are embedded. Though this is not a prerequisite, spheroid culture in low adhesion round-bottom wells is often used as well. As a load-bearing tissue, mechanical stimulation regimes like compression or hydrostatic pressure have been applied to improve physiological tissue architecture.

Interestingly, also tensile forces were used to improve cartilage maturation. For this, a special shaped agarose mold for self-assembly was created. It is comprised of a rectangular shape (8 × 13 mm) with four holes integrated to allow compatibility with tensile loading device. Seven to eight million articular chondrocytes per construct of bovine or human origin were left in the mold for self-assembly and mounted into the tensile loading device after 7 days and strained 15% with daily increases of 4–5% from days 8–12, from then on held constant until day 28. With this they were able to generate native tissue-like tensile properties in this neocartilage constructs [35].

Take-Home Messages
- Aggregates represent a shapeless multicellular entity formed via self-assembly.
- Spheroids are mainly defined by their spherical shape.
- Organoids are 3D in vitro tissues that possess structural and functional key properties of an organ at a micro- to millimeter scale.
- Aggregate culture was developed as early as in 1906, when the hanging drop method was established, using ESCs and embryoid body (EB) culture techniques to study intercellular signaling of early embryonic development.
- Multicellular aggregates are formed by self-organization utilizing cell-repellent plates, coating surfaces with non-adhesive properties, or hanging drop technique.
- Single cell suspensions undergo a three-phase process of spheroid formation: cell–cell affinity, reorganization and self-assembly, and lastly compaction.
- While growing, heterogenous gradients of oxygenation, nutrient supply, and hence cellular proliferation and necrosis are created within the aggregates.
- 3D cultures exhibit increased viability, proliferation, and differentiation potential as well as paracrine activity.
- Scaffold-free approaches have shown significant improvement regarding cell viability, growth, differentiation as well as towards related metabolism, genotype, and phenotype.
- Organoids are widely used for development and disease modeling, precision medicine, toxicology studies, and regenerative medicine.
- Organoids enable tissue culture for personalized medicine recapitulating organotypic structure and function.

(continued)

- Dynamic cultivation systems such as stirred tank bioreactors, perfusion bioreactors, horizontal or orbital shaking platforms demonstrate to have substantial positive impact on the cellular behavior compared to conventional static cultivation.

References

1. Li Y, Xu C, Ma T. In vitro organogenesis from pluripotent stem cells. Organogenesis. 2014:159–63.
2. Min S, Kim S, Cho S-W. Gastrointestinal tract modeling using organoids engineered with cellular and microbiota niches. Exp Mol Med. 2020;52:227–37.
3. Fair KL, Colquhoun J, Hannan NR. Intestinal organoids for modelling intestinal development and disease. Philos Trans R Soc Lond B Biol Sci. 2018;373:1750.
4. Barkauskas CE, Chung M-I, Fioret B, Gao X, Katsura H, Hogan BL. Lung organoids: current uses and future promise. Development. 2017;144:986–97.
5. Qian X, Song H, Ming G. Brain organoids: advances, applications and challenges. Development. 2019;146:166074.
6. Mansour AA, Schafer ST, Gage FH. Cellular complexity in brain organoids: current progress and unsolved issues. Semin Cell Dev Biol. 2020;14:115.
7. Prior N, Inacio P, Huch M. Liver organoids: from basic research to therapeutic applications. Gut. 2019;68:2228–37.
8. Miyoshi T, Hiratsuka K, Saiz EG, Morizane R. Kidney organoids in translational medicine: disease modeling and regenerative medicine. Dev Dyn. 2020;249(1):34–45.
9. Kruczek K, Swaroop A. Pluripotent stem cell-derived retinal organoids for disease modeling and development of therapies. Stem Cells. 2020;38:1206.
10. Corrò C, Novellasdemunt L, Li VS. A brief history of organoids. Am J Physiol-Cell Physiol. 2020;319:4.
11. Clevers H. Modeling development and disease with organoids. Cell. 2016;165(7):1586–97.
12. Lehmann R, Lee C, Shugart E, Benedetti MR, Charo A, Gartner Z, Hogan B, Knoblich J, Nelson C, Wilson K. Human organoids: a new dimension in cell biology. Mol Biol Cell. 2019;30(10):1129–37.
13. Bratt-Leal AM, Carpenedo RL, McDevitt TC. Engineering the embryoid body microenvironment to direct embryonic stem cell differentiation. Biotechnol Prog. 2009;25(1):43–51.
14. Simian M, Bissell MJ. Organoids: a historical perspective of thinking in three dimensions. J Cell Biol. 2017;216(1):31–40.
15. Antoni D, Burckel H, Josset E, Noel G. Three-dimensional cell culture: a breakthrough in vivo. Int J Mol Sci. 2015;16:5517–27.
16. Egger D, Tripisciano C, Weber V, Dominici M, Kasper C. Dynamic cultivation of mesenchymal stem cell aggregates. Bioengineering. 2018;5:2.
17. Cheng N, Chen S, Li J, Young T. Short-term spheroid formation enhances the regenerative capacity of adipose-derived stem cells by promoting stemness, angiogenesis, and chemotaxis. Stem Cells Transl Med. 2013;2(8):584–94.
18. Lin R-Z, Chang H-Y. Recent advances in three-dimensional multicellular spheroid culture for biomedical research. Biotechnol J. 2008:1172–84.

19. Tsai A-C, Liu Y, Yuan X, Ma T. Compaction, fusion, and functional activation of three-dimensional human mesenchymal stem cell aggregate. Tissue Eng. 2015;21:1705–19.
20. Ahmed S, Chauhan VM, Ghaemmaghami AM, Aylott JW. New generation of bioreactors that advance extracellular matrix modelling and tissue engineering. Biotechnol Lett. 2019;41:1–25.
21. Frith JE, Thomson B, Genever PG. Dynamic three-dimensional culture methods enhance mesenchymal stem cell properties and increase therapeutic potential. Tissue Eng. 2009;16(4):735–49.
22. Atac B, Wagner I, Horland R, Lauster R, Marx U, Tonevitsky AG, Azar RP, Lindner G. Skin and hair on-a-chip: in vitro skin models versus ex vivo tissue maintenance with dynamic perfusion. Lab Chip. 2013;13:18.
23. Egger D, Schwedhelm I, Hansmann J, Kasper C. Hypoxic three-dimensional scaffold-free aggregate cultivation of mesenchymal stem cells in a stirred tank reactor. Bioengineering. 2017;4:2.
24. Elanzew A, Sommer A, Pusch-Klein A, Brüstle O, Haupt S. A reproducible and versatile system for the dynamic expansion of human pluripotent stem cells in suspension. Biotechnol J. 2015;10:1589.
25. Gordon S, Daneshian M, Bouwstra J, Caloni F, Constant S, Davies D, Dandekar G, et al. Non-animal models of epithelial barriers (skin, intestine and lung) in research, industrial applications and regulatory toxicology. *ALTEX*. 2015;32(4):327–78.
26. Jung Y, Ji H, Chen Z, Chan HF, Atchison L, Klitzman B, Truskey G, Leong KW. Scaffold-free, human mesenchymal stem cell-based tissue engineered blood vessels. Sci Rep. 2015;5:15116.
27. Kelm JM, Ehler E, Nielsen LK, Schlatter S, Perriard J-C, Fussenegger M. Design of artificial myocardial microtissues. Tissue Eng. 2004;10:201.
28. Kelm JM, Djonov V, Hoerstrup SP, Guenter CI, Ittner LM, Greve F, et al. Tissue-transplant fusion and vascularization of myocardial microtissues and macrotissues implanted into chicken embryos and rats. Tissue Eng. 2006;12:9.
29. Kelm JM, Lorber V, Snedeker JG, Schmidt D, Broggini-Tenzer A, Weisstanner M, et al. A novel concept for scaffold-free vessel tissue engineering: self-assembly of microtissue building blocks. J Biotechnol. 2010;148(1):46–55.
30. Kempf H, Olmer R, Kropp C, Haverich A, Ulrich M, Zweigerdt R. Controlling expansion and Cardiomyogenic differentiation of human pluripotent stem cells in scalable suspension culture. Stem Cell Rep. 2014;3(6):1132–46.
31. Klingenhutz AJ, Gourronc FA, Chaly A, Wadkins DA, Burand AJ, Markan KR, Idiga SO, Wu M, Potthoff MJ, Ankrum JA. Scaffold-free generation of uniform adipose spheroids for metabolism research and drug discovery. Sci Rep. 2018;8:523.
32. Koenig L, Ramme A, Faust D, Lauster R, Marx U. Production of human induced pluripotent stem cell-derived cortical neurospheres in the DASbox® mini bioreactor system. Eppendorf AG Appl. 2018:364.
33. Lancaster MA, Renner M, Martin C-A, Wenzel D, Bicknell LS, et al. Cerebral organoids model human brain development and microcephaly. Nature. 2013:373–9.
34. Lee EJ, Park SJ, Kang SK, kim G-H, Kang H-J, Lee S-W, Jeon HB, Kim H-S. Spherical bullet formation via E-cadherin promotes therapeutic potency of mesenchymal stem cells derived from human umbilical cord blood for myocardial infarction. Mol Ther. 2012;20(7):1424–33.
35. Lee K, huwe LW, Paschos N, Aryaei A, Gegg CA, Hu JC, Athanasiou KA. Tension stimulation drives tissue formation in scaffold-free systems. Nat Mater. 2017;16:864–73.
36. Lin XZ, Geng X, Drewell C, Hübner J, Li Z, Zhang Y, Xue M, Marx U, Li B. Repeated dose multi-drug testing using a microfluidic chip-based coculture of human liver and kidney proximal tubules equivalents. Sci Rep. 2020;10:8879.

37. Lindner G, Horland R, Wagner I, Atac B, Lauster R. De novo formation and ultra-structural characterization of a fiber-producing human hair follicle equivalent in vitro. J Biotechnol. 2011;152(3):108–12.
38. Maschmeyer I, Lorenz AK, Schimek K, hasenberg T, Ramme AP, Hübner J, Lindner M, Drewell C, Bauer S, Thomas A, Sambo NS, Lauster R, Marx U. A four-organ-chip for interconnected long-term co-culture of human intestine, liver, skin and kidney equivalents. Lab Chip. 2015;15:2688–99.
39. Urich E, Patsch C, Aigner S, Graf M, Iacone R, Freskgard P-O. Multicellular self-assembled spheroidal model of the blood brain barrier. Sci Rep. 2013;3:1500.
40. Weng Y-S, Chang S-F, Shih M-C, Tseng S-H, Lai C-H. Scaffold-free liver-on-a-chip with multiscale organotypic cultures. Adv Mater. 2017;29:36.
41. Hildebrandt C, Büth H, Thielecke H. A scaffold-free in vitro model for osteogenesis of human mesenchymal stem cells. Tissue Cell. 2011;43(2):91–100.
42. Wimmer RA, Leopoldi A, Aichinger N, Wick B, Hantusch M, Novatchkova J, Taubenschmid, et al. Human blood vessel organoids as a model of diabetic vasculopathy. Nature. 2019;565 (7740):505–10.
43. Friedrich J, Seidel C, Ebner R, Kunz-Schughart LA. Spheroid-based drug screen: considerations and practical approach. Nat Protoc. 2009;4(3):309–24.
44. Abbasalizadeh S, Larijani MR, Samadian A, Baharvand H. Bioprocess development for mass production of size-controlled human pluripotent stem cell aggregates in stirred suspension bioreactor. Tissue Eng. 2012;18(11):831–51.
45. L'Heureux N, Pâquet S, Labbé R, Germain L, Auger FA. A completely biological tissue-engineered human blood vessel. FASEB J. 1998;12:47–56.
46. L'Heureux N, Dusserre N, Konig G, Victor B, Keire P, Wight TN, et al. Human tissue-engineered blood vessels for adult arterial revascularization. Nat Med. 2006;12:361–5.
47. TissUse GmbH. Emulating Human Biology. https://www.tissuse.com/en/humimic/ (n.d.). Accessed 30 July 2020.

Microfluidic Systems and Organ (Human) on a Chip

8

Janina Bahnemann, Anton Enders, and Steffen Winkler

Contents

J. Bahnemann (✉) · A. Enders · S. Winkler
Institute of Technical Chemistry, Leibniz University Hannover, Hannover, Germany
e-mail: jbahnemann@iftc.uni-hannover.de

© Springer Nature Switzerland AG 2021
C. Kasper et al. (eds.), *Basic Concepts on 3D Cell Culture*, Learning Materials in Biosciences,
https://doi.org/10.1007/978-3-030-66749-8_8

What You Will Learn in This Chapter

In the previous chapters we learned how cells are cultivated in 3D and how the surrounding gel matrix is optimized. However, to achieve even higher physiologically relevant cell culture conditions, the surrounding environment must be controlled by emerging microfluidic systems. Thus, in the first part of this chapter we will learn about the tremendous benefits of microfluidic devices, their fabrication, and finally their implementation in novel and highly controlled biological and cell culture applications.

On this basis, the second part of this chapter will focus on the complete control of biochemical and biomechanical cell culture parameters, which results in sophisticated organ-on-a-chip systems. You will learn how the blood–tissue barrier and the minimal functional unit of an organ are reconstructed to mimic specific organ functions. Finally, the combination of several different organ-on-a-chip systems results in the so-called human-on-a-chip systems. Although these systems are still in its infancy, we will elaborate on first design concepts and point out their future role in drug development processes in industry.

8.1 Introduction to Microfluidics

Microfluidics is an emerging interdisciplinary field that holds great promise for applications in such diverse realms as chemistry, biochemistry, and biological applications. Generally speaking, microfluidic systems involve the precise control of minute amounts of fluids (measured in the microliter scale) within complex channel systems at low flow rates. Because most microfluidic applications currently lie in chemical or biological analysis, these systems are also frequently referred to as Micro Total Analysis System (μTAS). The primary advantages of microfluidic systems over more traditional methods are comparatively small sample and reagent volumes, shorter analysis time, and lower cost.

But these are not the only advantages—other emerging applications in this field are leveraging the possibility of including several different analytic steps (i.e., mixing, diluting, and separating) in parallel within a single microfluidic system. Such systems are colloquially known as "Lab-on-a-Chip," and they hold great promise across a wide range of applications in the fields of biology and biochemistry—for example, by facilitating protein crystallization, cell growth analysis, or cultivation optimization [1]. And microfluidic devices are also increasingly being deployed within the biomedical field—for example, in the form of ready-to-use diagnostic systems for the so-called point-of-care diagnosis [2].

8.2   Characteristics of the "Microfluidic Environment"

Most microfluidic systems exploit specific characteristics of the unique environment that is created within these miniaturized conditions. As discussed below, such micro-environments are characterized by three important elements: (1) a small physical size, (2) an increased surface-to-volume ratio, and (3) a comparatively stable laminar flow.

The main advantage created by the small physical size of these systems is their decreased inner volume. This decrease in systemic volume allows for significantly lessened reagent and sample consumption as compared to more traditional methods. In addition, when using cells or particles, this small physical size also allows the user to influence every single cell more directly, via the introduction of nutrients, hormones, or other signaling molecules.

Surface-to-volume ratio also naturally increases as a direct function of decreasing channel size within these systems. This, in turn, has three corollary effects: First, a correlative increase in heat transfer across all parts of the system, which results in higher control and less dispersion. Second, a correlative increase in gas exchange. And third, a decrease in diversity within and across fluids within the system.[1]

Finally, due to the small physical dimensions and high surface-to-volume ratios typically found in microfluidic systems, fluid flow is typically dominated by viscous forces. As a result, the characteristic "Reynolds number" (defined as $Re = \rho v d / \eta$, where ρ is the density of the fluid, v is the velocity, d is the hydraulic diameter, and η is the dynamic viscosity of the fluid) in these systems is typically well below the threshold value of 2300—which means that the flow rate in microfluidic systems can be considered to be highly laminar. Unlike turbulent flow, where fluidic streamlines often cross (envision pouring milk into coffee or the turbulent water channels created by the movement of a ship's propeller), fluidic streamlines move "side-by-side" within a laminar flow (Fig. 8.1). As a result of this even and parallel motion, fluidic mixing is only caused by diffusion *at the interface of these streams*. This feature allows the architect of a microfluidic system to

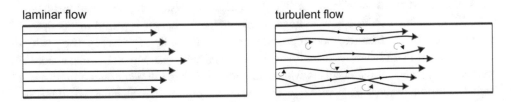

laminar flow   turbulent flow

Fig. 8.1   Schematic visualization of fluidic streamlines within laminar vs. turbulent flows

---

[1]Conversely, it should be noted that adsorption effects on channel walls also tend to increase. This is worth mentioning because adsorption can potentially lead to *unwanted* binding effects (e.g., with nonspecific proteins).

readily design channels that foster a stable concentration of different gradients [3] within several different units. If mixing is thereafter desired, then micromixer channels can also be introduced (see Chap. 4.3).

There are multiple ways to propel fluids in microfluidic systems. In some applications, like paper microfluidics, the effects of water interacting with the chip material are sufficient to propel the fluid forward. However, in more complex microfluidic systems, active pumps are needed to achieve suitable flow rates.

Some applications, like the Organ-on-a-Chip or Human-on-a-Chip designs discussed later in this chapter, rely on pulse-free slow fluid motion, which can be created best using pressure driven pumps. Pressure driven pumps use pressurized gas to force fluids into microfluidic channels. Other devices do not rely on pulse-free motion and can be used with more common peristaltic pumps, where a rotor is squeezing a pump tube in a revolving motion, thereby forcing the fluid inside the tube to move. Another common type of pump is the syringe pump, which provides accurate flow control and nearly pulse-free flow by pushing or pulling on a syringe filled with fluid but is also limited by the volume of the syringe used. Some designs even include a way to move fluids inside of the microfluidic system itself. In these devices, elastic properties of the chip material are used as a kind of membrane, which can be actuated using pressurized gas or force applied from the outside of the device. Therefore, the fluid motion can be controlled on the microfluidic device directly.

8.3   Microfluidic Fabrication Techniques

Perhaps not surprisingly, the roots of the field of microfluidics lie in the microelectronics industry. The very first microfluidic systems were etched into silicon wafers, using traditional electronics manufacturing principles. But silicon is expensive and opaque (which is a critical limitation with respect to designing biological experiments). As a result, biologists eager to take advantage of this emerging technology quickly began exploring glass and plastics as fabrication materials. The soft elastomer polydimethylsiloxane (PDMS) is widely used today in biological applications, where it is prized for its optical transparency, elastic features, biocompatibility, and permeability to gases [3]. The production process of PDMS systems usually involves several steps (see Fig. 8.2). First, a mold is created. The layout of the channels within this mold is printed directly onto a mask via a high-resolution printer. This mold is then used to outline the channel positions on photoresistant material on a silicon wafer, using a photolithographic process. The cured material is left protruding off the silicon. This is called the "master." The PDMS system is thereafter cast using the master and cured for 2 h at 60 °C. After the curing process is complete, the PDMS stamp is removed. The PDMS system now has the channel structures etched at the bottom and can be adhered to another flat sheet of PDMS, glass, or other materials, in order to seal the channels. For more complex three-dimensional channel systems, multiple PDMS stamps can also be adhered to one another [4].

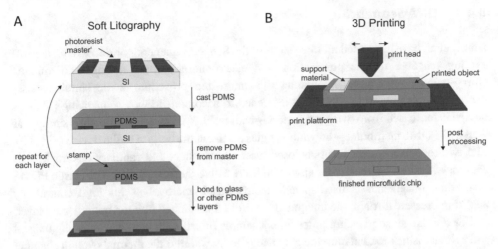

Fig. 8.2 Common fabrication techniques for creating microfluidic systems: (**a**) soft lithography (PDMS molding) and (**b**) 3D printing

In recent years, 3D printing has become more widely available, with the advent and popularization of new printing techniques, greater material selection, cheaper printers, and more refined printing resolution. As a result, 3D printing—which can frequently be done even more quickly, efficiently, and cost-effective than PDMS molding (Fig. 8.2)—has emerged as a highly promising alternative method for manufacturing microfluidic systems. Having said that, a note of caution is warranted: the materials used in "traditional" 3D printing applications (such as mechanical engineering and design) frequently are not biocompatible or chemically stable when exposed to solvents. This imposes a substantial limitation in the current use of 3D printing to create microfluidic systems within the biochemical and medical fields. Furthermore, even with recent advances in 3D printing techniques, current printing resolution still cannot compete with the smallest channel sizes achieved by using PDMS molding and photolithography. Nevertheless, the authors anticipate that the march of technological progress on this front will result in 3D printing becoming the "preferred" production mechanism for microfluidic systems in the near- to medium-term future, as biocompatible and chemically stable materials are being released [5, 6] and print resolutions continue to improve.

8.4   Overview of Biological and Cell Culture Applications

Numerous microfluidic devices have already been manufactured for use in various biological and cell culture applications [7]. Below, we highlight just a few of these devices and also briefly discuss the advantages they offer over more traditional methods.

8.4.1   DNA Sequencing

Blazej et al. have integrated all three steps of the Sanger sequencing protocol (e.g., thermal cycling reaction, sample purification, and electrophoresis) onto a single microfluidic chip—thereby effectively automating this multi-step procedure while simultaneously decreasing the amount of DNA that is needed to accomplish DNA sequencing [8]. This device is fabricated from two glass wafers—with features etched into both glass surfaces as well as a PDMS membrane—and another glass wafer at the bottom of the device (Fig. 8.3). The two top glass wafers form enclosed reaction chambers and capillary electrophoresis chambers, and the second wafer also includes resistive thermal probes. The elastic PDMS membrane underneath is used to actuate integrated microvalves for fluid control, by applying pressure through the integrated manifold lines etched into the bottom glass layer.

For the Sanger sequencing procedure, a sample mixed with both a sequencing reagent and primers is first loaded onto the chip and then moved into the thermal cycling chamber via the integrated valves. This mixture is then thermal cycled 35 times between 95 °C and 60 °C, at which point complementary DNA strands are synthesized. Each of these synthesized DNA strands is called Sanger extension fragments. Afterwards, the sample is pumped into the purification chamber, where a small polyacrylamide gel with single strand of DNA (complementary to the 3′ end) captures these Sanger extension fragments. Salts,

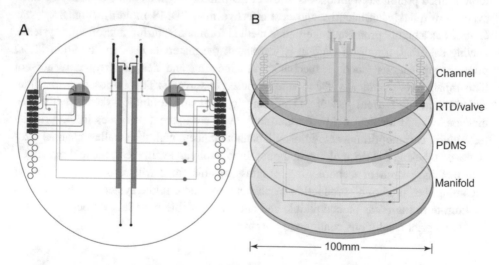

Fig. 8.3 An integrated nanoliter-scale DNA bioprocessor for Sanger sequencing. (**a**) schematic visualization of the top glass plates—illustrating the channel system for sample loading (red), the reaction chamber (orange), and the purification chamber and capillary electrophoresis (black); (**b**) overview of the entire system—showing the top glass plate which contains the channel system; the second glass plate which contains the thermistors and valve system; the elastic PDMS membrane which is used for valve actuation; and the bottom glass plate which contains the manifold channel system used to actuate the PDMS membrane via pressurized air. Blazej et al. [8] Copyright (2006) National Academy of Sciences, U.S.A.

primer, and excess sample DNA are then passed through the gel and removed. After capture, the Sanger extension fragments are released by heating the purification chamber and thereafter moved into the separation capillary where capillary electrophoresis is performed. At the end of the capillary, the four-color sequence data is collected on a radial scanner. Through this mechanism, Blazej et al. were able to achieve a sequence accuracy of 99% using only 1 fmol of DNA template while demonstrating long read lengths suitable for the de novo sequencing of complex genomes. The integration of all three steps onto one system also reduces both the manual labor required to operate the system and the volumes of reagents and DNA needed to conduct the analysis.

Aborn et al. have also developed a system for high-throughput DNA sequencing by parallelizing the electrophoresis step of the Sanger method. In this approach, polyacryl-amide gel is pumped into 384 enclosed separation lanes [9]. Afterwards, the prepared Sanger fragments are loaded onto the lanes, and electrophoresis is performed. At the end of these lanes, a multi-line laser is used to excite the DNA and a scanning detector is used to capture the fluorescence data. Using two of these 384-lane plates in parallel, they were able to parallelize the cleaning and loading steps for a 768-lane complete system—which can sequence more than four million bases per day. This system highlights the great potential that microfluidic systems hold for parallelizing operations on a micro-scale.

8.4.2 Point-of-Care Diagnostics

The possibility of miniaturization inherent to microfluidics not only allows for new and improved methods of analytical procedures, but also facilitates new applications in medical diagnostics. The field of point-of-care (POC) diagnostics uses microfluidic systems to develop small, low cost, and self-contained devices that provide analysis directly at the patient instead of relying on analytical laboratories. Major advantages include not just the reduced timeframe to complete a diagnosis, but also the greater independence from infrastructure—which is particularly beneficial in developing countries and/or in disaster situations.

The iSTAT device by Abbott is one of the oldest and most successful commercially available POC devices. This handheld, battery-powered microfluidic device is used for detection of blood chemistries (like potassium, sodium, and glucose), coagulation, and cardiac markers. The analyzer handles drops of whole blood (approximately 100 μL) without sample preparation and uses internal calibration. The calibration reagents are integrated on disposable plastic test cartridges containing an air bladder for fluid movement, a small channel system, and a silicon microchip. Micro-fabricated thin-film electrodes on this microchip coated with ionophores or enzymes are used for detection of various analytes. For fluid movement of the sample and calibration fluids, an electric motor in the handheld device presses on the air bladder on the test cartridge. The handheld nature and power-independence of the device makes it an excellent example of a practical POC device [10].

Another commercially available POC device is the PIMA CD4 counter, manufactured by Abbott. This device is used to count CD4 cells in AIDS/HIV patients—a disease especially prevalent in developing countries. The device employs static image analysis for cell counting. The sample (25 μL of capillary blood) is pumped into a disposable cartridge with dry sealed reagents. In the first compartment, fluorescent anti-CD3 and anti-CD4 antibodies bind to their respective target cells. Then the sample is transferred to a detection area, where the stained cells are imaged and analyzed using image analysis algorithms. The whole process takes just 20 min and can be performed with minimal training in a small desktop system [10]. Although there are a few established systems already on the market, the development of miniaturized, personalized, low-cost, and easy-to-use POC diagnostic systems continues to be a focus of research. In particular, the integration of suitable (bio)sensors into microfluidic systems has already increased sensor selectivity and sensitivity for the detection of specific biomarkers (such as proteins) [11]. In addition, integrated microfluidic POC devices offer the possibility of parallelizing and automating sample processing and analysis.

8.4.3   Handling of Suspension Cells

The ability to handle liquids with high precision in small volumes also makes microfluidics attractive for handling suspended cells. Often, these systems seek to combine several laborious steps into a single system to allow for parallelization, automation, and easier handling.

One example for these systems is an integrated system for fast dynamic quantitative analysis of the metabolism of mammalian suspension cells from a bioreactor [12]. This system combines the sample treatment, mixing, incubation, and sequential separation as well as media exchange in one system with two temperature zones to ensure physiological conditions (37 °C) as well as improved cell quenching at 4 °C. Therefore, the system does not only massively reduce the time and manual labor needed for sample processing, but also offers advantages by providing greater temperature control [13].

Another example of a microfluidic device for cell culture applications is a continuous system for transient transfection of Chinese hamster ovary (CHO) cells. The system aims to combine the necessary lab steps for DNA vector integration in CHO-K1 cells into one system: (1) mixing of DNA vector, chemical transfection reagent, and cell suspension, (2) incubation, and (3) separation of the cells (Fig. 8.4).

The first functional unit of the illustrated transfection system consists of an integrated micromixer. Enders et al. have demonstrated the efficiency of four different passive micromixers in a comparative analysis [14]. While the environment in microfluidic systems is typically laminar—which limits the mixing phenomena to slow diffusion—passive micromixers can be used to overcome the poor mixing in microfluidics by rearranging the flow and disturbing the parallel flow lines. One example for disturbing the flow lines is the popular Tesla-like mixing structure. This mixer splits the flow vertically and leads one

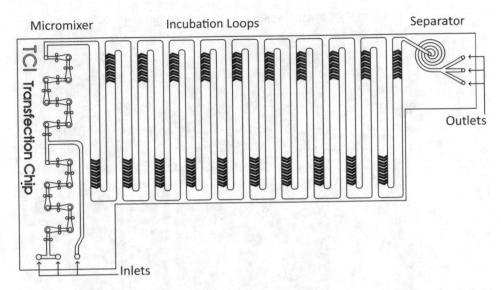

Fig. 8.4 Microfluidic system for transient cell transfection. This system combines cell and fluid mixing, incubation, and separation steps in one microfluidic device

half of the flow in a 180° turn to recombine with the other half head on. Conversely, the F-type mixers split the flow vertically and recombine it horizontally, which create an alternating pattern and decrease the distance for diffusion. Additional information on both passive and active micromixers can be found from Capretto et al. [15] or Nguyen et al. [16].

Enders et al. also used 3D printing to quickly fabricate the complex channel structures of the different micromixers. Figure 8.5 illustrates the printing workflow. Additionally, the 3D printing workflow is shown in our video, which can be viewed via the following link (https://youtu.be/Wc4gjoxfhOw) and the QR code in Fig. 8.5. The complete printing process is dependent on the model dimensions, but generally takes about 2–4 h in total.

Another functional unit of the microfluidic transfection system is the separation of cells following the transfection. The aim is to preserve cell viability by separating the cells from the toxic transfection reagent. Microfluidic separation systems can focus particles of various sizes at specific points inside a microfluidic channel. An example of a simple separation system is a spiral separator. While particles converge to stable positions inside a channel in a laminar flow environment naturally, a channel with a rectangular cross-section only has two stable positions in a horizontal plane. When winding a rectangular channel into a spiral, a pressure difference is created between the flow at the outer wall and the inner wall of the spiraling channel. This leads to a new flow pattern orthogonal to the main flow direction (called "dean flow"), which leads the outer stable position in the channel to become unstable [17]. Thereafter, only one stable position—towards the inner wall of the channel—remains. The spiral separator is an ideal tool for focusing cells from suspension at this position.

Fig. 8.5 3D printing workflow used by Enders et al. (**A**) computer aided design (CAD); (**B**) the CAD file is sent directly to the 3D printer; (**C**) detailed view of the print head, which places the print model material and support material; (**D**) the channels are filled with support material during the printing process, which is removed during post-processing; (**E**) the finished 3D-printed micromixer

8.4.4 Analysis of Single Cells

Conventional analytical methods used for analyzing suspended cells (e.g., in cell culture or blood) frequently use the averaged results of a sample size of hundreds or thousands of cells. However, these methods do not consider cells as single individuals, so the information reflects only the average state of a population of cells. The use of microfluidic platforms opens the possibility of analyzing single cells.

Grünberger et al. have developed several systems consisting of very small bioreactors on microfluidic chips to perform various analysis on individual cells [18, 19]. The smallest system is a cell trap for a singular bacterial cell, which holds the cell in place while fresh medium flows around the cell. The group was able to trap *E. coli* cells and monitor the cell growth over several hours, showing constant division times and typical morphology (which indicates that the system exerted no inhibiting effects tarpon the cell) (Fig. 8.6). This system allows for live-cell imaging and analysis over extended periods of time in a perfectly controlled environment, which facilitates more granular analysis and observation of the response of a single bacteria to short term environmental fluctuations (e.g., in pH, temperature, etc.).

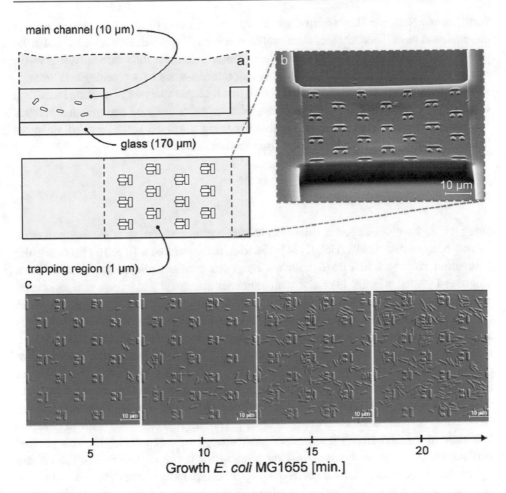

Fig. 8.6 Cell traps for single cell cultivation and studies by Grünberger et al. (**a**) schematic illustration of the complete chip and the trapping region, (**b**) microscopic view of the trapping region, (**c**) microscopic images demonstrating the growth of *E. coli* over time

Gao et al. have developed another microfluidic system for use in biological cell studies [20]. Unlike other cell analysis protocols, this system focuses on the analysis of a single cell. The team used human blood cells to conduct an analysis of intracellular constituents via a simple microfluidic device which consists of only four channels (sample input, buffer input, sample waste, and a capillary electrophoresis channel) leading to a single crossing point. First, the human blood cells (in suspension) were pipetted onto the chip and flowed into the channel system via hydrodynamic force. Then, a single cell was captured using electrophoretic means at the crossing point by applying a set of potentials at the end of the four channels. This cell was docked at the channel walls and then lysed by applying even higher voltages. This docked-lysing approach led to reduced dispersion of the released cell constituents. Next, capillary electrophoresis was performed in the capillary electrophoresis channel, and the cell constituents were analyzed using a fluorescence detector located at the end of the channel. Although the operation in this research was fully manually, it is worth noting that this process could also be easily automated—since only pipetting and voltage-switching are involved. It is also worth noting that multiple analyses of a single individual cell can provide researchers and/or doctors with far more nuanced and granular insights into the health of a patient's blood cells than using averaged results gained via more traditional homogenized samples taken from thousands of cells.

8.4.5 Parallel Cell Culture

Hung et al. have developed a microfluidic system for performing a parallel perfusion culture of mammalian cells (HeLa) [21]. The team manufactured a 10×10 grid of circular cultivation chambers in PDMS with an integrated gradient micromixer system. Each chamber featured a larger inlet and outlet channel at opposing sides, as well as several small perfusion channels surrounding the chamber. In experiments using this system, HeLa cells were first loaded in suspension and then left for 2 h to settle. Afterwards, these cells were fed using perfusion medium pumped at low flow rates through the perfusion feed, until the cells adapted to the new environment and cell growth stabilized at a normal rate (a time interval of 8 days). At that point, perfusion was stopped, and sample reagents were fed through the inlet channels. Using an integrated gradient micromixer system located in front of the inlet channels, ten different concentrations of reagents could be created, and ten wells could be utilized for each concentration. A Calcein AM cell assay with an observation time of 10 days (using fluorescence microscopy) was then deployed. This parallel perfusion culture system holds significant promise for future deployment in cell culture optimization and studies in tissue behavior. Again, comparatively small reagent and media volumes, as well as parallelization, enable more cost-effective assay methods within this microfluidic system when compared to more traditional methods.

Gómez-Sjöberg et al. have gone even farther and enhanced the idea of parallel cultivation of adherent mammalian cells by developing a fully automated cultivation system with 96 individually addressable cultivation chambers on a single chip (Fig. 8.7) [22]. Once again, every cultivation chamber was quite small—with a volume of just 40 nL—and the whole microfluidic chip was mounted on an automated microscope equipped with a motorized X-Y stage. During the cultivation process, each hour a phase contrast image was taken and then automatically analyzed. Even at the cell loading step, automatic cell counting was used to ensure that the exact same number of cells were placed into every cultivation chamber. Up to 16 reagents and culture media were connected to the system at the same time and mixed at different quantities.

Through this study, Gómez-Sjöberg et al. have amply demonstrated that a microfluidic system can simultaneously sustain proliferation while also stimulating the differentiation of human mesenchymal stem cells. By automating both the pumping of reagents/media and

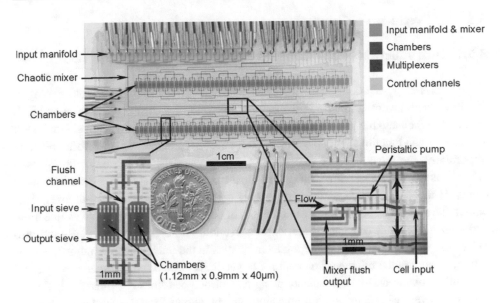

Fig. 8.7 A microfluidic cell culture array for perfusion culture, reprinted with permission from Gómez-Sjöberg et al. Copyright (2007) American Chemical Society

the microscopic imaging phases of their process, the group was able to implement complex and time-varying feeding and stimulation schedules while also taking time-lapse microscopic images of each cultivation chamber. As a result, they were able to study the effects of osteogenic differentiation factors on cell motility in a highly granular fashion.

A slightly different cultivation device was published by Siller et al. [23]: The 3D-printed cultivation vessel was used to co-cultivate endothelial and mesenchymal stem cells indirectly. A physical barrier was separating the cell types from one another, while medium was able to flow over that barrier. The 3D-printed material enabled phase contrast and fluorescence microscopy, which allowed for the observation of cell growth over time. These observations and further analysis revealed that endothelial cells form tubular-like structures when cultivated alongside mesenchymal stem cells, a feature that can be considered angiogenic. In addition, this study demonstrated that the 3D-printed material is biocompatible and thus suitable for the development of individual cell culture vessels.

The foregoing examples—which represent only a small sampling of some recent applications of microfluidic systems that have been deployed in recent years—illustrate the varied and numerous advantages that these systems can offer to researchers in the fields of biotechnology and bioengineering: parallelization, automation, small sample and reagent volumes, and more direct sample control chief among them.

8.5   "Organ-on-a-Chip"

8.5.1   Introduction to the Concept of the "Organ-on-a-Chip" (OoC)

"Organ-on-a-chip" (OoC) systems represent one of the most promising biotechnologies that have been invented to date in the field of microfluidics. Just as the lab-on-a-chip is intended to facilitate the miniaturization and automatization of lab procedures, OoCs seek to mimic organ functions by deploying 3D cell culture techniques on a microfluidic chip. Importantly, the focus in these systems is *not* the reconstruction of complete organs— rather, it is merely the development of "minimal functional units" (MFUs) that accurately represent the core essence of these organs as it relates to experimental purposes. This innovative approach encompasses a wide range of concepts—including tissue engineering, hydrogel integration, cell integration, cell cultivation, and cell differentiation—and also makes use of a wide array of emerging technologies in the field, including pumps, valves, and the design of the microenvironment for targeted 3D culture formation and cultivation. Currently, advanced OoC systems are using established cell lines or primary cells. However, by its very nature, the microfluidic environment allows for the precise control of cell culture conditions and spatial- and time-dependent differentiation of stem cells within OoCs. As a result, adult or induced pluripotent stem cells are assuming an increasingly prominent role in this research field [24].

The highest potential for OoC technology lies in the pharmaceutical industry—particularly within the context of drug development. At present, the long pipeline of new drug development includes (without limitation) the synthesis of chemical compounds; high-throughput screenings for biological activity and toxicity by using in vitro enzymatic or cell-based assays; pre-clinical investigation of pharmacokinetics and dynamics by in vivo animal experiments; and, finally, three separate clinical phases culminating in studies featuring thousands of patients. Unfortunately, the second and third clinical testing phases (which occur relatively late in this chain) are the single most expensive steps for drug development—and they are also characterized by the highest failure rates. As a result, *ex ante* predictions of compound activity, toxicity, and other key benchmarks for drug candidates derived from data realized using current in vitro and in vivo techniques are notoriously unreliable.

OoC holds tremendous promise as a tool to help bridge the expensive and time-consuming gap that currently separates the pre-clinical and clinical phases [25]. It balances the advantages of high reliability of in vitro techniques with the higher physiological relevance of in vivo parameters by including novel in vivo-like parameters into the system (Fig. 8.8). These in vivo-like parameters are additional biomechanical parameters that more closely mimic actual in vivo conditions. For instance, mechanical stimulations have been demonstrated to influence cell behavior—in particular, the process of cell differentiation. This is because most, if not all, organs are exposed to (at least small) mechanical forces. For instance, a lung-on-a-chip system may help to stimulate the formation of the lung epithelial cell barrier by using a vacuum-induced cell strain. And even static organs are exposed to

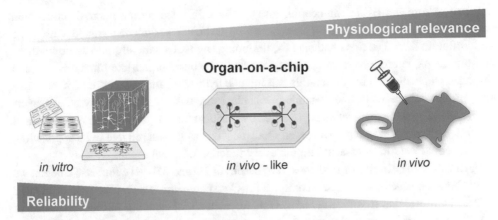

Fig. 8.8 Organ-on-a-chip systems balance the advantages of high reliability of in vitro techniques with the higher physiological relevance of in vivo parameters by including novel in vivo-like parameters

the basic mechanical forces caused by the liquid flow of the blood within the organ tissue. OoCs have revealed a fluid flow shear stress that alters gene expression—to cite one example, this stress has been found to influence vasculature diseases [26]. Similar to the vascular system, the flow additionally allows the permanent supply and removal of anabolites and metabolites, thereby creating a constant pH value and a constant distribution of oxygen, medium, and drugs in a physiological liquid-to-cell ratio. These in vivo-like parameters make OoCs superior to the standard in vitro 3D cell culture. Furthermore, animal experiments can also be complemented, or even entirely supplanted, by OoCs. The applicability of animal models to human patients is notoriously limited. Indeed, this disjunction currently represents a major bottleneck for drug development. But OoCs have substantially better predictive capabilities, because they combine in vivo-like advantages with *human* cells. As a result, tests for compound properties (such as liver toxicity or skin irritations) can be more reliably conducted using human liver-on-a-chip or skin-on-a-chip systems. And in contrast to using animals as black boxes, the processes in OoC devices can also be electrochemically or optically monitored by integrated sensors or microscopic/spectroscopic techniques, resulting in a high data output that finally can be multiplied by automation and high-throughput screenings.

8.5.2 Engineering the Organ-on-a-Chip Microenvironment

Engineering in vivo-like complexity within the organ-on-a-chip microenvironment is highly dependent on an interdisciplinary combination of tissue engineering technology and microfluidic knowledge. This is accomplished in an iterative process, where the requirements for survival and function of cells and cell cultures—investigated and defined by cell culture and tissue engineering research—are fulfilled by the microstructures and control of the chip, which is in turn based on technical knowledge gained within the microfluidic field. The first challenge for designing any OoC is therefore to ascertain these requirements, in order to reconstruct and maintain the basic organ-like functions of a cell culture. Depending on the organ, this can include hydrogels, membranes, electrical fields, gas or nutrient gradients, and dynamic mechanical stimulation, among many other aspects [27]. Defining these parameters is governed by a critical manta: "as much as necessary and as little as possible." In other words, an OoC must be usable and reliable in its application, but it should *not* (unnecessarily) mimic every function of a real organ.

An OoC can be built from either ex vivo tissues or 2D and 3D cell cultures. To begin, we will focus on general steps and techniques for OoC engineering using the type of 2D and 3D cell cultures that have been used in OoC research so far—saving further discussion of OoCs of important organs for later on.

Engineering a successful OoC is strongly dependent on the selection of a suitable cell source. Primary cells, "immortalized" cell lines, and various kinds of stem cells are all popular choices. Primary cells are unmodified mature cells, and, perhaps not surprisingly, their major advantage is their similarity to in vivo tissue. However, they also suffer from several distinct disadvantages—including ethical and practical difficulties associated with isolating them from other animal or human tissue; limited resources and lifetime; a comparatively difficult cultivation; and an unfortunate tendency to alternate gene expression and the loss of function after a few weeks of cultivation. As a consequence, many OoC platforms instead make use of the so-called immortalized cell lines that are derived from primary cells via a process of chemical or viral modification. Immortalized cells are easier to cultivate and also exhibit superior growth when compared with primary cells. Nonetheless, like the primary cells from which they are derived, immortalized cells also tend to display genotypic and/or phenotypic alternations as the culture matures. Stem cells are perhaps the most promising of all, because they offer the possibility of differentiation into complex in vivo-like tissues. Thus, OoC platforms could manage a targeted manipulation of cell differentiation to create a specific assembly of different cell types creating a more physiological, tissue-like cell culture. Depending on their origin, stem cells can be classified into embryonic stem cells (ESCs), adult stem cells, or induced pluripotent stem cells (iPSCs). iPSCs are of particular interest to researchers, because at least in principle they can be achieved by reprogramming skin cells isolated from any person—thereby enabling truly personalized OoC models.

After the appropriate cell type(s) have been selected and their requirements are ascertained and defined, the microenvironment can then be engineered. One of the primary engineering challenges faced by every OoC model is the reconstruction of the blood–tissue barrier, where the microfluidic nutrient flow represents the blood and the 3D cell culture organization represents the tissue [28]. This barrier is a central biological principle found in every organ, which allows the perfusion and nutrient supply of the 3D culture. Depending on the organ, additional barriers must be reconstructed to manage separation of, e.g., urine in kidneys; bile in liver; food in the gut; air in the lungs; etc. Several basic techniques for

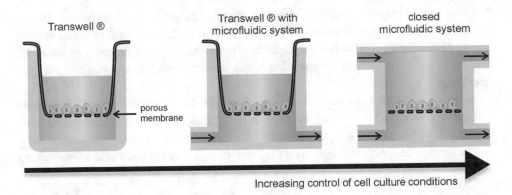

Transswell ®

Transwell ® with
microfluidic system

closed
microfluidic system

porous
membrane

Increasing control of cell culture conditions

Fig. 8.9 The advances of membrane-based organ-on-a-chip systems. Early Transwell® systems included a permeable membrane, but no flow control (left). In contrast, many current membrane-based OoCs make use of microfluidic platforms for Transwell® integration (middle) or complete displacement (right), to allow liquid flow that results in an increased control of cell culture conditions

creating these interfaces inside the microfluidic chip have been developed. The use of porous membranes is very common.

The membrane allows the cultivation of 2D or 3D cell cultures on both sides: By manipulating the pore size, thickness, surface properties, elasticity, and other parameters, the membrane can be adapted to fulfill all of the complex functions that are required in a barrier (including selective permeability, cell attachment, cell migration, and cell alignment). Porous membranes also offer comparatively good transparency to facilitate microscopic observation. The so-called Transwell® is one of the simplest and best-known examples of a membrane-based OoC design (Fig. 8.9). Initially it was plugged in standard well-plates, but, with advances in microfluidics, it has been combined with microfluidic devices to enable a continuous flow. Today, many membrane-based OoCs replace the barrier by integrated membranes—allowing a closed design accompanied with markedly higher control of cell culture conditions. Unfortunately, the use of membranes in 3D cultures is limited by the fact that they cannot (presently) be freely modeled in all three dimensions.

As discussed in Chap. 5, 3D cultivation can be accomplished by using hydrogels to form a physiological extracellular matrix (ECM) for cell encapsulation. One approach to hydrogel integration—derived from microfluidic chips fabrication technology—is soft lithography. In this process, a non-polymerized gel is squeezed into a reusable template, which is then removed after gel polymerization occurs. Another approach is selective photopolymerization using photomasks with subsequent removal of non-polymerized gel. However, as 3D microstructures become more complicated, the challenges associated with using either of these methods quickly escalates. At a micro-scale, adhesion and capillary forces can be used to exclusively trap the gel in specific channels without blocking neighboring channels for fluid flow. Porous or degradable chip materials enable the construction of permeable or disappearing barriers between hydrogel and fluid channels that will not be overgrown and blocked by cells due to the constant perfusion [29]. In combination with advancing 3D printing techniques, these channels can be fabricated in all dimensions of space—allowing a well-defined perfusion of the hydrogel. Likewise, 3D printing can directly be used for microstructuring hydrogel using bioprinting technologies (see Chap. 9).

Many current OoCs are membrane-based and thus do not require a sophisticated 3D tissue microstructure for fulfilling the requirements of their specific application. But OoC research is still in its infancy—and with the increasing success of OoCs spurring on ever-more-complex designs, the demands on OoC design will only continue to increase moving forward.

8.5.3   Reconstructing the "Minimal Functional Unit"

As discussed above, every organ contains its own "minimal functional unit" (MFU), which fulfills the basic functions of the organ. The overall aim of any OoC is to reconstruct this unit—and *only* this unit. In contrast to 2D, 3D, or even organoid cultures, the MFU encompasses biomechanical functions such as (by way of example) contraction, dilation, resorption, filtration, and excretion. However, mimicking *all* functions of the MFU is extremely challenging, and, as a result, current OoCs typically focus on only a few of them. Nevertheless, researchers all over the world are currently reconstructing organ functions for nearly every human organ—even the brain. Below, we briefly discuss how MFUs of the liver and kidney are mimicked by existing OoC systems.

Liver Because of the central role that it plays in metabolizing drugs within the human body, the liver is a common focus for OoC systems. The MFU of the liver is the liver sinusoid. The sinusoid is a capillary that combines oxygen-rich blood from the artery with nutrient-rich blood from the portal vein. Liver sinusoidal endothelial cells (LSECs), Kupffer cells, hepatic stellate cells, and hepatocytes [30] all act together to form a porous barrier between the capillary and the bile duct, where hepatocytes clean the blood by scavenging and metabolizing toxic substances. This blood–tissue barrier has most commonly been mimicked by using integration of permeable membranes to allow cultivation of LSECs and hepatocytes on opposing sides [30, 31]. Other approaches involve nanostructures which allow diffusion of nutrients and removal of waste products [32]. However, many liver-on-a-chip devices omit this barrier and instead merely contain 2D cell monolayers or 3D spheroids. This is because for studying the toxicity and metabolism of drugs, sinusoid-like structures are generally not absolutely mandatory. Nevertheless, the liver sinusoid *is* essential for observing urea secretion function—and as a result its inclusion in a liver-on-a-chip OoC is critical for studying pharmacokinetic and/or pharmacodynamic drug behavior [33].

Kidney The kidney is also of special interest for drug testing, since it eliminates xenobiotics and is highly involved in regulating blood pressure. The MFU of the kidney is the nephron, which consists of the glomerulus, proximal tubule, and the loop of Henle. Current microfluidic devices serve a specific purpose—such as characterizing drug transport and nephron toxicity [34]—and as a result they tend to focus on replicating glomeruli or tubule structures. The main function of the glomerulus is the filtration of blood. Similar to a liver-on-chip OoC, endothelia cells and glomeruli-specific cell types (such as podocytes) have been cultured on each side of a permeable membrane in kidney-on-a-chip OoCs. Furthermore, mechanical influences like pulsation of the renal blood flow have also been reconstructed via vacuum-driven deformation of the membrane [35]. The main function of the proximal tubule is the reabsorption of solute and fluid [36], which has similarly been reconstructed in several tubule-on-a-chip systems by a membrane-based design [34, 37, 38]. However, the tubule has also been designed by 3D cell culture techniques as well, using hollow fibers or bioprinting [39, 40]. The 3D gel matrix is formed in a tubular structure allowing cells to be cultured inside or on the surface of the matrix.

8.6 "Human-on-a-Chip" and "Disease-on-a-Chip"

8.6.1 The Principles of Multi-Organ-on-a-Chip

The concept of combining several different OoCs in a single chip is called human-on-a-chip (HoC), body-on-a-chip, multi-organ-on-a-chip, or micro-cell culture analogs (μCCAs) (Fig. 8.10). Although several multi-organ-on-a-chip devices have been created to date [41–44], no complete HoC has been successfully developed yet, because the reconstruction of all organs and their interactions remains a subject of active and ongoing research. Nevertheless, the concepts for multi-organ-on-a-chips and HoCs are essentially the same: A HoC must be a stable system where OoCs can interact and communicate with each other, while unwanted fluctuations that potentially lead to non-physiological functions are prevented or minimized.

To engineer such a stable system, OoCs must firstly be scaled down by using either allometric scaling or residence-time based scaling approach [45]. Allometric scaling miniaturizes organs relative to each other according to their scaling factors. But miniaturization of a human body from kilograms to grams, or even milligrams, does not follow a trivial linear (isometric) down-scaling of all organs. This is because different characteristics of an organ—such as mass, metabolization, blood volume, or oxygen consumption—all follow different scaling factors [46]. To illustrate this concept, consider cells that are already relatively rare within a normal, full-sized human body: An isometrically scaled HoC would include only a few leukocytes, and they would be inhomogeneously distributed within the system. In consequence, they could not fulfill their functions in all compartments of the chip. Moreover, scaling always changes ratios of physical quantities. Thus, diffusion has a significantly higher impact in small HoC devices than in a real full-sized body. One of

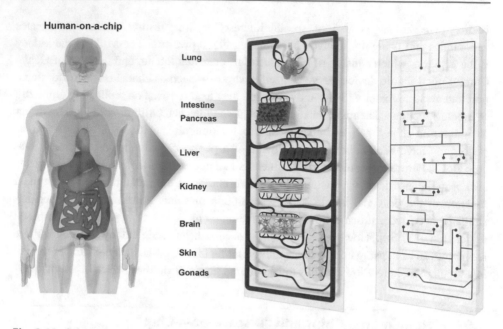

Fig. 8.10 Schematic presentation of a "human-on-a-chip" (HoC). In a HoC several different organ-on-a-chip systems are combined to simulate the physiology of the human body

the main issues with engineering HoCs for pharmacokinetic drug testing is that scaling does not change enzyme/protein affinities. As a result, a liver-on-a-chip may not produce a physiological relevant concentration of drug metabolites—which further cannot activate or block their targets or off-targets in a lung-on-a-chip. Residence-time-based scaling seeks to tackle this issue by determining the concentrations drugs and metabolites inside the OoCs and then replicating physiologically relevant drug concentrations.

OoCs must also be connected in the correct way within a HoC. For instance, nutrient uptake starts with absorption in the gut, which is first delivered to the liver via the portal vein, then modified and released into the bloodstream, and then partially excreted by the kidney. As a result, the chip must include interconnecting channels that mimic all of these different main connection pathways (e.g., the bloodstream, the urine stream, etc.). Depending on the device, organs can be created inside the chip simultaneously in a universal medium, separately with a subsequent combination, or even partially combined by connecting and counterbalancing two OoCs first [47]. The last two approaches offer the possibility of OoC exchange. In other words, a single impaired OoC can be readily replaced and does not necessarily lead to dysfunction within the larger system. Furthermore, culturing cells of different OoCs with a single medium is challenging, because a single common medium cannot satisfy the specific needs of every cell type, and the device may not maintain its supposed function. In consequence, a flexible approach had been proposed, where organs can be cultured separately and connected as soon as they are needed for an experiment [47].

8.6.2 "Human-on-a-Chip"

To date, multi-organ-on-a-chip devices predominantly integrate OoCs which are of particular interest for pharmaceutical industry—such as the liver, kidney, gut, skin, and lungs. These organs are crucial for pharmacokinetic investigations of adsorption, distribution, metabolism, excretion, and toxicity (ADME-Tox) of pharmaceutical compounds within the human body.

The ADME-Tox behavior of a compound is currently described via animal experiments, and, as a result, the highest potential for HoCs using human cells can be seen in ADME-Tox characteristics that are highly species-dependent—such as metabolism and toxicity. The metabolism of a compound, particularly within the liver, often differs widely between animals and human—leading to very different by-products with different levels of toxicity. While in vitro cell cultures of human liver cells are already being used to investigate and assess these by-products for basic liver toxicity, detecting and understanding possible toxic effects within other organs remains a project for future studies. HoC technology promises to open that door to researchers. But many other parameters of the ADME-Tox process can also be monitored within HoCs: In contrast to animals, a HoC can be comprehensively modified and adapted to allow the integration of sensor spots or live cell imaging to monitor changes in the concentrations of pH, oxygen, or metabolites of interest, as well as cell morphologies and cell culture properties.

One important caveat: HoCs are still in development, and at this point, it remains highly challenging to attempt to mimic all important organ functions for reliable ADME-Tox predictions (and thereby truly replace animal testing). As a result, from a practical perspective, in the context of drug testing HoCs are perhaps best viewed as a gradual complement to animal testing rather than a complete replacement looming on the horizon. This complementation can be obtained by comparing physiologically based pharmacokinetic (PBPK) models gained from animal experiments, animal-on-a-chip (AoC) systems, and HoC systems (Fig. 8.11) [48]. The AoC acts as a linker for in vivo to in vitro extrapolation, where its PBPK is optimized to match the PBPK from animal experiments. Following this, the AoC PBPK can be extrapolated across species by comparing it to the PBPK of the corresponding HoC. Thus, in principle at least, a HoC with the same design should fit the AoC PBPK model, when a drug has the same behavior in animal and in human. In turn, differences in these models could help to identify deviant pharmacokinetic characteristics of a drug and thereby help to prevent drug failure in later (and more costly) clinical stages.

8.6.3 "Disease-on-a-Chip"

Engineering a disease-on-a-chip (DoC) is another promising application for tissue engineers. In principle, a DoC is nothing more than a slightly modified OoC. Nevertheless, in practice, the DoC is even more complex—because it needs to reconstruct disease processes as well as organ functions. This can include the use of genetically modified cells; the targeted integration or reconstruction of dysfunctional tissues; or the integration

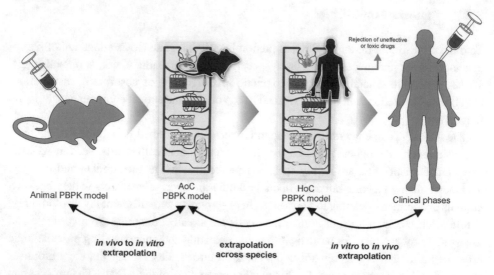

Fig. 8.11 The future role of human-on-a-chip systems in drug development. Pharmacokinetic and pharmacodynamic data of animal experiments are summarized to a physiologically based pharmacokinetic model (PBPK). This model will be used for in vivo to in vitro extrapolation to develop a reliable animal-on-a-chip model (AoC). In turn, the PBPK of the AoC will be used for the extrapolation to human and for the development of a HoC. Finally, the HoC PBPK model will help to reject ineffective or toxic drugs before entering costly and time-consuming clinical phases

of tissue-unrelated cell types or even organisms. A simple example of reconstructing a disease is the thrombosis-on-a-chip [49]. A thrombosis includes the agglomeration of thrombocytes at the vascular wall that form a thrombus, which then blocks the blood flow in the vein. First, a vessel-on-a-chip system is constructed by forming hydrogel into a chip including channels for liquid flow, where endothelial cells have been seeded and cultured. Then, the disease element is integrated into the system by inducing thrombosis in a blood flow inside the channels using calcium chloride ($CaCl_2$).

Like OoCs, DoCs can serve as in vivo-like platforms to reduce animal experiments and offer higher reliabilities by using human cells. But DoCs can also be used to investigate the activity of drug candidates and observe how they counteract specific diseases. In contrast to using animal models, the comparative accessibility of a chip creates new opportunities for researchers to observe and understand disease processes—particularly on the microcellular or intercellular level. A cancer-on-a-chip platform is a prominent example of reconstructing and investigating tumor processes. The altered metabolism and the resulting tumor environment have been observed by culturing tumor cells in a 3D gel matrix with constant perfusion [50]. The artificial tumor created pH and oxygen gradients that led to different gene expression profiles dependent on the location in the tumor.

Cancer-on-a-chip platforms are also of particular interest for personalized DoCs, because tumor cells develop due to several random mutations in the cell genome, and every tumor can therefore evidence a slightly different behavior—even when the main tumor growth-inducing mutations are nearly identical. As a result, personalized platforms are favored by integrating tumor cells of a single patient into DoCs for subsequent drug screenings to rapidly establish patient-specific therapies.

Although research into OoCs, HoCs, and DoCs remains at a very early stage at this time of this publication, these microfluidic 3D cell culture platforms are developing rapidly. Companies are investing extensively into promising new biotechnologies, and OoC start-ups are popping up all over the world—highlighting the tremendous promise that these systems offer to revolutionize the fields of cell cultivation, tissue engineering, and medical research in general.

Take-Home Message
- Microfluidics involve the precise control of minute amounts of fluids within complex channel systems at low flow rates.
- "Lab-on-a-chip" systems are microfluidic chips that allow the miniaturization and automation of lab procedures such as DNA sequencing, point-of-care diagnostics, cell transfection, single cell analysis, and parallel cell culture.
- Microfluidic chips are commonly fabricated via soft lithography or emerging 3D printing techniques.
- An "organ-on-a-chip" (OoC) is a cell culture system typically inside a microfluidic chip that mimics the minimal functional unit of a specific organ.
- A "human-on-a-chip" (HoC) combines several OoC systems to mimic human physiology.
- A "disease-on-a-chip" (DoC) is an OoC or HoC system which additionally mimics a pathophysiologic process.
- OoCs, HoCs, and DoCs all hold immense promise for revolutionizing drug testing in the pharmaceutical industry.

References

1. Whitesides GM. The origins and the future of microfluidics. Nature. 2006;442:368–73. https://doi.org/10.1038/nature05058.
2. Pandey CM, Augustine S, Kumar S, et al. Microfluidics based point-of-care diagnostics. Biotechnol J. 2018;13:1700047. https://doi.org/10.1002/biot.201700047.
3. Weibel DB, Whitesides GM. Applications of microfluidics in chemical biology. Curr Opin Chem Biol. 2006;10:584–91. https://doi.org/10.1016/j.cbpa.2006.10.016.
4. Whitesides GM, Ostuni E, Takayama S, Jiang X, Ingber DE. Soft lithography in biology and biochemistry. Annu Rev Biomed Eng. 2001;3(1):335–73.
5. Siller IG, Enders A, Steinwedel T, et al. Real-time live-cell imaging technology enables high-throughput screening to verify in vitro biocompatibility of 3D printed materials. Materials (Basel). 2019;12:2125. https://doi.org/10.3390/ma12132125.

6. Siller IG, Enders A, Gellermann P, et al. Characterization of a customized 3D-printed cell culture system using clear, translucent acrylate that enables optical online monitoring. Biomed Mater. 2020;15:055007. https://doi.org/10.1088/1748-605X/ab8e97.

7. Yeo LY, Chang H-C, Chan PPY, et al. Microfluidic devices for bioapplications. Small. 2011;7:12–48. https://doi.org/10.1002/smll.201000946.

8. Blazej RG, Kumaresan P, Mathies RA. Microfabricated bioprocessor for integrated nanoliter-scale sanger DNA sequencing. Proc Natl Acad Sci U S A. 2006;103:7240–5. https://doi.org/10.1073/pnas.0602476103.

9. Aborn JH, El-Difrawy SA, Novotny M, et al. A 768-lane microfabricated system for high-throughput DNA sequencing. Lab Chip. 2005;5:669–74. https://doi.org/10.1039/b501104c.

10. Chin CD, Linder V, Sia SK. Commercialization of microfluidic point-of-care diagnostic devices. Lab Chip. 2012;12:2118–34. https://doi.org/10.1039/c2lc21204h.

11. Arshavsky-Graham S, Enders A, Ackerman S, et al. 3D-printed microfluidics integrated with optical nanostructured porous aptasensors for protein detection. Microchim Acta. 2021;188:67. https://doi.org/10.1007/s00604-021-04725-0

12. Bahnemann J, Rajabi N, Fuge G, et al. A new integrated lab-on-a-chip system for fast dynamic study of mammalian cells under physiological conditions in bioreactor. Cell. 2013;2:349–60. https://doi.org/10.3390/cells2020349.

13. Rajabi N, Bahnemann J, Tzeng T-N, et al. Lab-on-a-chip for cell perturbation, lysis, and efficient separation of sub-cellular components in a continuous flow mode. Sensors Actuators A Phys. 2014;215:136–43. https://doi.org/10.1016/j.sna.2013.12.019.

14. Enders A, Siller IG, Urmann K, et al. 3D printed microfluidic mixers-a comparative study on mixing unit performances. Small. 2019;15:e1804326. https://doi.org/10.1002/smll.201804326.

15. Capretto L, Cheng W, Hill M, et al. Micromixing within microfluidic devices. Top Curr Chem. 2011;304:27–68. https://doi.org/10.1007/128_2011_150.

16. Nguyen N-T, Wu Z. Micromixers—a review. J Micromech Microeng. 2005;15:R1–R16. https://doi.org/10.1088/0960-1317/15/2/R01.

17. Di Carlo D. Inertial microfluidics. Lab Chip. 2009;9:3038–46. https://doi.org/10.1039/b912547g.

18. Probst C, Grünberger A, Wiechert W, et al. Polydimethylsiloxane (PDMS) sub-Micron traps for single-cell analysis of bacteria. Micromachines (Basel). 2013;4:357–69. https://doi.org/10.3390/mi4040357.

19. Grünberger A, Wiechert W, Kohlheyer D. Single-cell microfluidics: opportunity for bioprocess development. Curr Opin Biotechnol. 2014;29:15–23. https://doi.org/10.1016/j.copbio.2014.02.008.

20. Gao J, Yin X-F, Fang Z-L. Integration of single cell injection, cell lysis, separation and detection of intracellular constituents on a microfluidic chip. Lab Chip. 2004;4:47–52. https://doi.org/10.1039/b310552k.

21. Hung PJ, Lee PJ, Sabounchi P, et al. Continuous perfusion microfluidic cell culture array for high-throughput cell-based assays. Biotechnol Bioeng. 2005;89:1–8. https://doi.org/10.1002/bit.20289.

22. Gómez-Sjöberg R, Leyrat AA, Pirone DM, et al. Versatile, fully automated, microfluidic cell culture system. Anal Chem. 2007;79:8557–63. https://doi.org/10.1021/ac071311w.

23. Siller IG, Epping N-M, Lavrentieva A, et al. Customizable 3D-printed (co-)cultivation systems for in vitro study of angiogenesis. Materials (Basel). 2020;13:4920. https://doi.org/10.3390/ma13194290.

24. Zhang J, Wei X, Zeng R, et al. Stem cell culture and differentiation in microfluidic devices toward organ-on-a-chip. Future Sci OA. 2017;3:FSO187. https://doi.org/10.4155/fsoa-2016-0091.

25. Marx U, Andersson TB, Bahinski A, et al. Biology-inspired microphysiological system approaches to solve the prediction dilemma of substance testing. ALTEX. 2016;33:272–321. https://doi.org/10.14573/altex.1603161.
26. Cecchi E, Giglioli C, Valente S, et al. Role of hemodynamic shear stress in cardiovascular disease. Atherosclerosis. 2011;214:249–56. https://doi.org/10.1016/j.atherosclerosis.2010.09.008.
27. Zhang B, Radisic M. Organ-on-a-chip devices advance to market. Lab Chip. 2017;17:2395–420. https://doi.org/10.1039/c6lc01554a.
28. Zhang B, Korolj A, Lai BFL, et al. Advances in organ-on-a-chip engineering. Nat Rev Mater. 2018;3:257–78. https://doi.org/10.1038/s41578-018-0034-7.
29. Zhang B, Montgomery M, Chamberlain MD, et al. Biodegradable scaffold with built-in vasculature for organ-on-a-chip engineering and direct surgical anastomosis. Nat Mater. 2016;15:669. https://doi.org/10.1038/nmat4570.
30. Du Y, Li N, Yang H, et al. Mimicking liver sinusoidal structures and functions using a 3D-configured microfluidic chip. Lab Chip. 2017;17:782–94. https://doi.org/10.1039/c6lc01374k.
31. Domansky K, Inman W, Serdy J, et al. Perfused multiwell plate for 3D liver tissue engineering. Lab Chip. 2010;10:51–8. https://doi.org/10.1039/b913221j.
32. Lee PJ, Hung PJ, Lee LP. An artificial liver sinusoid with a microfluidic endothelial-like barrier for primary hepatocyte culture. Biotechnol Bioeng. 2007;97:1340–6. https://doi.org/10.1002/bit.21360.
33. Hedaya MA. Basic pharmacokinetics, Pharmacy education series. 2nd ed. Hoboken: CRC Press; 2012.
34. Jang K-J, Mehr AP, Hamilton GA, et al. Human kidney proximal tubule-on-a-chip for drug transport and nephrotoxicity assessment. Integr Biol (Camb). 2013;5:1119–29. https://doi.org/10.1039/c3ib40049b.
35. Musah S, Mammoto A, Ferrante TC, et al. Mature induced-pluripotent-stem-cell-derived human podocytes reconstitute kidney glomerular-capillary-wall function on a chip. Nat Biomed Eng. 2017;1:0069. https://doi.org/10.1038/s41551-017-0069.
36. Weinberg E, Kaazempur-Mofrad M, Borenstein J. Concept and computational design for a bioartificial nephron-on-a-chip. Int J Artif Organs. 2008;31:508–14. https://doi.org/10.1177/039139880803100606.
37. Sciancalepore AG, Sallustio F, Girardo S, et al. A bioartificial renal tubule device embedding human renal stem/progenitor cells. PLoS One. 2014;9:e87496. https://doi.org/10.1371/journal.pone.0087496.
38. Kim S, LesherPerez SC, Kim BCC, et al. Pharmacokinetic profile that reduces nephrotoxicity of gentamicin in a perfused kidney-on-a-chip. Biofabrication. 2016;8:15021. https://doi.org/10.1088/1758-5090/8/1/015021.
39. Ng CP, Zhuang Y, Lin AWH, et al. A fibrin-based tissue-engineered renal proximal tubule for bioartificial kidney devices: development, characterization and in vitro transport study. Int J Tissue Eng. 2013;2013:1–10. https://doi.org/10.1155/2013/319476.
40. Homan KA, Kolesky DB, Skylar-Scott MA, et al. Bioprinting of 3D convoluted renal proximal tubules on perfusable chips. Sci Rep. 2016;6:34845. https://doi.org/10.1038/srep34845.
41. Maschmeyer I, Lorenz AK, Schimek K, et al. A four-organ-chip for interconnected long-term co-culture of human intestine, liver, skin and kidney equivalents. Lab Chip. 2015;15:2688–99. https://doi.org/10.1039/c5lc00392j.
42. Skardal A, Murphy SV, Devarasetty M, et al. Multi-tissue interactions in an integrated three-tissue organ-on-a-chip platform. Sci Rep. 2017;7:8837. https://doi.org/10.1038/s41598-017-08879-x.

43. Miller PG, Shuler ML. Design and demonstration of a pumpless 14 compartment microphysiological system. Biotechnol Bioeng. 2016;113:2213–27. https://doi.org/10.1002/bit. 25989.
44. Tsamandouras N, Chen WLK, Edington CD, et al. Integrated gut and liver microphysiological systems for quantitative in vitro pharmacokinetic studies. AAPS J. 2017;19:1499–512. https://doi.org/10.1208/s12248-017-0122-4.
45. Abaci HE, Shuler ML. Human-on-a-chip design strategies and principles for physiologically based pharmacokinetics/pharmacodynamics modeling. Integr Biol (Camb). 2015;7:383–91. https://doi.org/10.1039/c4ib00292j.
46. Wikswo JP, Block FE, Cliffel DE, et al. Engineering challenges for instrumenting and controlling integrated organ-on-chip systems. IEEE Trans Biomed Eng. 2013;60:682–90. https://doi.org/10. 1109/TBME.2013.2244891.
47. Rogal J, Probst C, Loskill P. Integration concepts for multi-organ chips: how to maintain flexibility?! Future Sci OA. 2017;3:FSO180. https://doi.org/10.4155/fsoa-2016-0092.
48. Esch MB, King TL, Shuler ML. The role of body-on-a-chip devices in drug and toxicity studies. Annu Rev Biomed Eng. 2011;13:55–72. https://doi.org/10.1146/annurev-bioeng-071910- 124629.
49. Zhang YS, Davoudi F, Walch P, et al. Bioprinted thrombosis-on-a-chip. Lab Chip. 2016;16:4097–105. https://doi.org/10.1039/c6lc00380j.
50. Ayuso JM, Virumbrales-Munoz M, McMinn PH, et al. Tumor-on-a-chip: a microfluidic model to study cell response to environmental gradients. Lab Chip. 2019;19:3461–71. https://doi.org/10. 1039/c9lc00270g.

3D-Bioprinting

Daniela F. Duarte Campos and Andreas Blaeser

Contents

D. F. Duarte Campos
Department of Materials Science & Engineering, Stanford University, Stanford, CA, USA

AME – Institute of Applied Medical Engineering, RWTH Aachen University, Aachen, Germany

A. Blaeser (✉)
BioMedical Printing Technology, TU Darmstadt, Darmstadt, Germany

Centre for Synthetic Biology, TU Darmstadt, Darmstadt, Germany
e-mail: blaeser@idd.tu-darmstadt.de

© Springer Nature Switzerland AG 2021
C. Kasper et al. (eds.), *Basic Concepts on 3D Cell Culture*, Learning Materials in Biosciences,
https://doi.org/10.1007/978-3-030-66749-8_9

What You Will Learn in This Chapter
Bioprinting is a digital biofabrication method that is applied to generate complex and heterogenic 3D cell culture systems and living tissues. It adopts the concept of additive manufacturing to replicate the macroscopic anatomy of the tissue of interest by layer-wise printing of living cells embedded in a hydrogel matrix. The printed precursor structures are cultivated for several weeks in dedicated bioreactors, on-chip platforms, or host organisms to mature and form biofunctional tissue units.

This chapter will give you a detailed overview of 3D-bioprinting technology, its challenges as well as its current and future applications. Following a brief introduction, this chapter provides a step-by-step guide through individual operations throughout the bioprinting process chain.

Beginning with the generation of printable 3D-datasets obtained from medical imaging modalities, such as computed tomography (CT) and magnetic resonance imaging (MRI), and computer-aided design (CAD), this chapter discusses the formulation of bioinks from natural and synthetic origins. Holistically, this chapter discusses different bioprinting methods and mechanisms, ranging from stereolithography and microextrusion to laser-assisted and microvalve-based drop-on-demand printing.

This chapter closes with an overview of current and future applications of 3D-bioprinting for clinical purposes, such as implants, tissue substitutes, in vitro models, and organs-on-a-chip.

Learning Objectives
- Definition of 3D-bioprinting technology and its applications.
- Individual elements of the 3D-bioprinting process chain.
- Widely applied bioprinting mechanisms, methods, and strategies.
- Application of bioprinting technology in medicine.
- Tissue design considerations and post-printing maturation.

9.1 From Tissue Engineering to 3D-Bioprinting

Bioprinting describes an additive manufacturing technology for the fabrication of living tissue that evolved in the late 1990s and early 2000s. Cell-laden hydrogels are deposited layer-by-layer to create a viable precursor structure that can be further matured to form functional, living tissue. Its development started in a time with rapid progress in related technological and biomedical fields of research, such as the market launch of 3D-printing systems and the evolution of tissue engineering.

Pursuing the goal to substitute, restore, or maintain the function of damaged organs, tissue engineering combines biological, medical, and engineering approaches to recreate living tissue [1]. In brief, living cells can be embedded in a hydrogel matrix, seeded on top of a 3D-structure, or cast in a mold that resembles the tissue geometry, and cultured in the presence of nutrients and soluble growth factors. In 1999, the first tissue-engineered structure, a bladder, was successfully implanted by Anthony Atala [2]. Until today, the described molding and seeding methods are successfully applied for the fabrication of planar, cylindrical, or hollow tissues, such as heart valves and blood vessels [3, 4].

In order to automize and upscale the manual, labor-intensive tissue engineering procedures, the idea of bioprinting was born. Cells and matrix materials, referred to as bioink (Sect. 9.2.2), would no longer need to be cast or seeded manually, but could be deposited automatically utilizing robotics and automated fluid handling systems. Initially, modified desktop printers were used for this task. The inkjet cartridges were cleaned from ink, refilled with cell suspension, and moved horizontally to print drops of bioink on a substrate mounted to a leverageable, vertical axis [5]. Even though bioprinting technology rapidly evolved since then, the basic setup is still comparable to current systems, which comprise a three- to six-axis robotic platform and multiple printer heads. However, the printer heads, the applied printing mechanism, as well as the printing strategies have become much more versatile and specialized since then. A detailed overview is given in Sect. 9.3.

Besides automation, tissue engineering was thought to further benefit from additional advantages promised by 3D-printing technology, such as spatially controlled, reproducible deposition of different bioinks as well as the fabrication of complex and patient-individualized 3D-geometries. Ultimately, bioprinting could even be a promising path towards the fabrication of thick, vascularized tissues and complex organs, which is the holy grail in tissue engineering. How far these promises could be kept, and which challenges remain, will be subject of the following chapters. Starting with the required printing preparations (Sect. 9.2), the bioprinting process (Sect. 9.3) will be elucidated. This book chapter closes with a summary of current and future applications of this inspiring technology (Sect. 9.4).

> **Questions**
> 1. How is the term Tissue Engineering defined?
> 2. What were/are the motivations to apply bioprinting technology?

9.2   3D-Bioprinting Process Preparation

The 3D-bioprinting process can be subdivided into three phases: print preparation, printing procedure, and post-printing tissue maturation (Fig. 9.1). This chapter describes the preparatory work that needs to be conducted in the first phase before running the

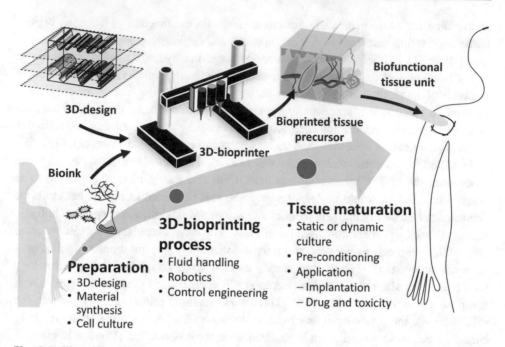

Fig. 9.1 Illustration of the bioprinting process chain comprising three phases: print preparation, printing procedure, and post-printing tissue maturation. The image is reproduced from the *Biofabrication and 3D-bioprinting* lecture notes with permission of Prof. Blaeser, Institute for BioMedical Printing Technology, Technical University Darmstadt

bioprinting process. First, the 3D-geometry that is supposed to be replicated must be defined and transformed into printable data (Sect. 9.2.1). Next, the bioink and its components (living cells and hydrogels) must be prepared, mixed, and filled into the bioprinting cartridges. Both steps will be explained in more detail in the following subchapters.

9.2.1 Acquisition of Printable 3D-Data Sets

The generation of printable 3D-data sets in bioprinting is very similar to the well-established slicing processes of conventional 3D printing software. Still, the origin of the 3D data as well as the complexity of the reproduced object might differ. This book chapter will therefore not elucidate the applied algorithms in detail, but rather give a broad overview of the required steps and focus on the differences and peculiarities that must be considered in bioprinting. For further reading and information about slicing algorithms please refer to Pandey as well as Donghong and coworkers [6, 7].

First, a virtual model of the desired 3D-structure is generated by CAD, 3D-scanning, or medical imaging (e.g. CT, MRI). Next, the files are converted and stored in a format, which can be interpreted by a slicing tool. The STL-format, for instance, is well-established and compatible with most slicing software. The format describes the surface area of the 3D-geometry as a triangular mesh generated by tessellation algorithms [8–10]. Each triangular facet is described by an outward-directed normal vector and three-point coordinates [8].

Especially in bioprinting, 3D-data sets might be obtained from medical imaging modalities rather than CAD. These must undergo additional segmentation steps before translation into a tessellated file format. CT or MRI typically create a 3D image of a certain region of the body, e.g. the chest. If only the geometry of the lung is supposed to be printed, it needs to be extracted from the surrounding tissue, e.g. bone, blood vessels, cartilage, fat, etc. Segmentation algorithms and software are commercially available and well described in the literature [11, 12].

Next, a so-called slicing algorithm is applied to virtually cut your 3D-object into a stack of thin slices with a defined thickness (Fig. 9.2). The slices resemble individual layers of the 3D-geometry that are subsequently printed to generate the final 3D-object. For each slice,

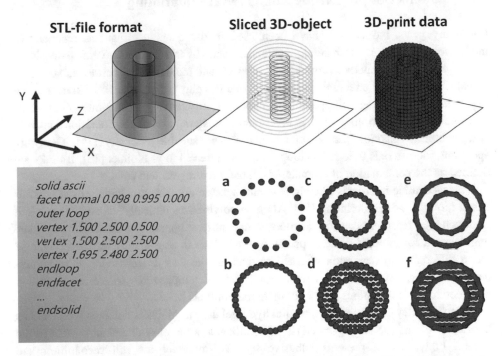

Fig. 9.2 Schematic illustration of the slicing process, required to generate 3D-printable data sets. Besides the three process steps (STL-File, Sliced 3D-object, 3D-print data) the general STL-file format is shown (left). Also, different print settings that can be selected using a drop-based slicing algorithm are depicted exemplary (**a–f**). The user can select to only print the outer contour (**a, b**), the outer and inner contour (**c**) or both contours as well as the filling (**d**). The contours and filling can either be printed using individual drops (**a–d**) or by microextrusion of plotted lines (**e, f**). The figure is reproduced from the user manual of the *SuperFill Software Suite 1.7* with permission of Black Drop Biodrucker GmbH

the intersecting points of an infinite horizontal plane at a defined distance from the coordinate system's origin and the sides of the triangular mesh are determined mathematically. Connecting the intersecting points, the outer- and inner contour of the two-dimensional slice can be calculated. The infill depicts the region that is encircled by the outer and (if present) inner contour.

Finally, the print path and machine commands are calculated for each layer and stored in a machine-readable format, e.g. G-code. These calculations can be individualized by the user and depend on the applied printing system and mechanism. For instance, it is possible to adjust the slicing height, structure, and density of the filling, as well as hardware settings, such as print head, extruder, and printing platform temperature, speed, feed rate and drop size.

9.2.2 Bioink Design Considerations for 3D-Bioprinting

For many years, biomaterials have been used in tissue engineering and regenerative medicine. Hydrogels of natural, synthetic, and hybrid origins were extensively investigated. These materials provide both physical and biological support to encapsulated cells [13]. Mechanics, structure, and matrix dynamics play an important role in instructing cell behavior, including cell spreading, migration, differentiation, and fate. The use of hydrogels in 3D bioprinting requires additional biomaterial design considerations, such as shear-thinning behavior, fast gelation kinetics, and suitable viscosity [14]. Generally speaking, biomaterials used in 3D bioprinting are referred to as bioinks [15]. Bioinks are defined as the combination of biomaterials, mostly hydrogels, and cells.

Polysaccharide hydrogels like alginate and agarose are the most widely used natural-based bioinks in 3D bioprinting [16]. Alginate polymerizes ionically via sodium-calcium ion exchange reaction occurring at room temperature. Alginate was used in bone tissue engineering, for example, for encapsulating osteoblasts in core–shell constructs. Agarose has a different polymerization mechanism than alginate, as its gelation is temperature-dependent. Agarose hydrogel is in a liquid state at temperatures above 32 °C and it polymerizes at temperatures below room temperature [17].

A further relevant example of a natural hydrogel used as bioink is collagen type I and its hybrids with alginate and agarose [18, 19]. Collagen is the most abundant protein in the human body, and it presents adhesive ligands important for cell recognition and remodeling. It has been used in combination with agarose for 3D bioprinting of bone and cartilage substitutes [20, 21], corneal tissue models [22], dental pulp [23], and cancer research [24].

From a panoply of synthetic polymers, polyethylene glycol (PEG) has been widely used in 3D bioprinting applications. As pure PEG is not suitable for bioprinting, due to its water-solubility, it has been combined with PEG diacrylate (PEGDA) [25] or even with other natural hydrogels like alginate [26].

Although 3D bioprinting has made great progress in the last decade, the choice of bioinks for the manufacturing process is still evolving. There is a demand for more advanced, adaptable bioinks that mimic more closely the native ECM of tissues and organs [13]. Polysaccharides, collagen, PEGDA, and other commonly used bioinks (gelatin, fibrin, PLA, PCL, etc.) are static matrices, which degrade over time. In the next years, it is expected that smart bioinks with responsive and adaptable biological, mechanical, and rheological characteristics will take over, as they can better replicate biological aspects occurring in native tissues [13, 27].

Questions
3. What are the individual transformation steps to acquire a printable 3D-data set from medical imaging (e.g. CT or MRI)?
4. Which are the two main components of a bioink?
5. Which natural hydrogels are mostly used as bioinks?

9.3   3D-Bioprinting Process

This chapter illustrates the 3D-bioprinting process. Following hardware initialization and a system-dependent cleaning protocol, the previously designed 3D-data and the freshly prepared bioink formulation are loaded into the printing system and the bioink cartridge, respectively. With the start of the printing procedure, one or multiple bioinks are printed layer-by-layer according to the previously defined print path (Sect. 9.2.1). The way how the materials are deposited and how individual layers are formed strongly depends on the applied bioprinting method (Sect. 9.3.1) and strategy (Sect. 9.3.2). The most frequently used technologies and strategies are described and discussed in the following subchapters.

9.3.1   3D-Bioprinting Methods and Mechanisms

The 3D-bioprinting method determines how discrete fractions of the prepared bioink are transferred from the cartridge or reservoir, where they are kept in a bulk volume, onto the substrate or a previously printed layer. Depending on the scale of these fractions, three different classes of bioprinting methods are distinguished: (1) layer-wise, (2) line-wise, and (3) drop-wise printing (Fig. 9.3) [28]. For instance in inkjet bioprinting, the smallest printable material fraction is a single drop, while projection micro-stereolithography immediately prints a full layer of your material. Each class might achieve material transfer through different mechanisms, which are explained in the next subchapters in greater detail.

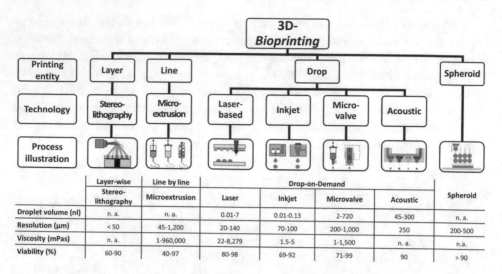

Fig. 9.3 Overview of the most commonly applied 3D-bioprinting technologies and their key parameters, reproduced from the *Biofabrication and 3D-bioprinting* lecture notes with permission of Prof. Blaeser, Institute for BioMedical Printing Technology, Technical University Darmstadt

To give a broad orientation to the reader, this subchapter informs on different printing characteristics and quality features, such as printing resolution (e.g. minimally achievable feature size, drop volume, or drop size), viscosity range of printable bioinks, and post-printing cell survival rates. Less frequently reported criteria, such as printing speed, drop frequency, and material throughput, are not taken into consideration. However, it is important to notice that most of the described quality features do not only depend on the applied printing method but are strongly influenced by the rheology of the bioink (e.g. viscosity, shear-thinning behavior, and surface tension), cell type, applied printing settings (e.g. pressure, nozzle size, temperature), and experience of the user. For these reasons, only average target corridors are given below. In individual cases, the characteristics described may be significantly exceeded or undercut.

9.3.1.1 Projection Stereolithography

Stereolithography is a widely spread 3D-printing method, where a light source, e.g. UV or laser light, is applied to polymerize a photo-curable resin, monomer solution, or hydrogel [29]. The light is directed towards a bath filled with the material to be processed either in a top-down or bottom-up orientation. In conventional stereolithography, individual spots of the precursor material are illuminated one-by-one with a focused beam. In projection stereolithography, the full layer to be printed is illuminated simultaneously, which speeds up the printing time (Fig. 9.4) [31].

Compared to other printing methods, stereolithography mainly stands out concerning printing accuracy. It enables printing feature sizes down to 20µm [29]. However, this

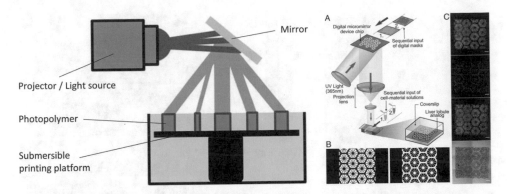

Fig. 9.4 Schematic representation of the projection stereolithography process (left), reproduced from the *Biofabrication and 3D-bioprinting* lecture notes with permission of Prof. Blaeser, Institute for BioMedical Printing Technology, Technical University Darmstadt. Ma and coworkers applied this method successfully to generate human iPSC-derived hepatic tissue models (right) [30]. (**a**)–(**c**) were reproduced with permission from PNAS

technology exhibits certain limitations when used for bioprinting. The liquid material precursor is often added as a photoinitiator or catalyst to initiate and accelerate the curing step [32]. These additives as well as the exposure to focused laser- or UV-light were shown to affect the viability and proliferation rate of bioprinted stem cells [33]. Under severe conditions, a reduction in the viability of more than 40% was observed. Even though the toxic and potentially carcinogenic effects of additive and UV-light exposure can be compensated over culture time [30], the risk of inducing long-term mutations in the cells currently limits the application of stereolithographic bioprinting to in vitro applications, such as organs-on-a-chip, and makes the fabrication of tissues for implantation questionable.

An additional limitation of stereolithography bioprinting is the maintenance of the material precursor in a bath, which can be relatively large compared to the printed structure. This situation forces the bioprinting process to require a large amount of dead volume, which is critical and inefficient when using bioinks. Moreover, the setup impedes the fabrication of multimaterial objects, which would require an exchange of material.

Nonetheless, this method has been applied successfully, for instance, in bioprinting tissue models for drug screening [30]. Furthermore, recent technological developments will potentially overcome the described multimaterial limitations. Changing the precursor material bath for a microfluidic chamber that can be filled, emptied, and exchanged with different fluids automatically, multimaterial printing of acellular constructs is already possible [34].

9.3.1.2 Microextrusion

Microextrusion is certainly the most widely used 3D-bioprinting method. It is a direct bioprinting technique that adopts the basic concept of fused filament fabrication (FFF) also

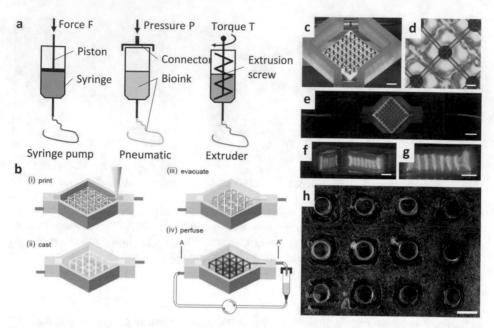

Fig. 9.5 Schematic representation of different fluid transport mechanisms applied in microextrusion bioprinting (**a**), reproduced with permission from the *Biofabrication and 3D-bioprinting* lecture notes of Prof. Blaeser, Institute for BioMedical Printing Technology, Technical University Darmstadt. The research group of Jennifer Lewis applied microextrusion bioprinting technology for the fabrication of thick vascularized tissues (**b–h**) [39], reproduced with permission from PNAS

known as fused layered modeling (FDM) to print single strands of bioink [35, 36]. The bioink is kept in a heatable/coolable reservoir and extruded through needles with adjustable diameters. The bioink can be squeezed through the nozzle by applying a mechanically actuated piston [37], pneumatic pressure [38], or a screw-extruder (Fig. 9.5) [36]. The feature size of printed lines ranges from 45 to 1250μm and strongly depends on the applied needle size, viscosity of the bioink, and applied extrusion mechanism [28, 37, 38, 40]. Its ability to precisely deposition a broad number of bioinks with viscosities ranging from 30 mPa.s to 960 kPa.s is the strongest advantage of this technology [40, 41]. Still, bioinks with viscosities in the mPa.s range tend to spread when extruded through the nozzle and compromise the bioprinting accuracy. In some studies, bioinks were therefore added additives or natural thickeners, such as nanocellulose, to enhance shape fidelity [42]. However, the use of bioinks with viscosities in the kPa.s range increases nozzle shear stress and printing related cell defects. The estimated cell survival rates following microextrusion may, therefore, vary from 40 to 95% [38, 41]. Recent developments, such as the application of microfluidic extrusion devices or printing of core–shell structures (Sect. 9.3.2.4) offer solutions to this challenge [43, 44].

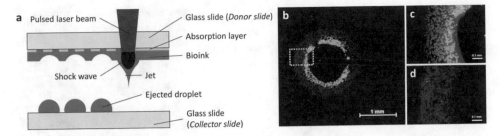

Fig. 9.6 Schematic illustration of the laser-based 3D-bioprinting process (**a**), adapted from Gruene et al. [46], and reproduced from the *Biofabrication and 3D-bioprinting* lecture notes with permission of Prof. Blaeser, Institute for BioMedical Printing Technology, Technical University Darmstadt. Guillotin and coworkers applied laser-assisted bioprinting technology for the fabrication of planar constructs made of fluorescently labeled HUVEC cell lines (Eahy926) in 1% alginate (**b–d**) [47], reproduced with permission from Elsevier

9.3.1.3 Laser-Assisted Bioprinting

Laser-assisted bioprinting describes a nozzle-free, digital printing modality that applies the mechanism of laser-induced forward transfer (LIFT) to deposit single drops of bioink [45]. Briefly, a focused laser beam is applied to eject a fraction of bioink from a donor slide towards an opposing collector (Fig. 9.6). The donor slide comprises a bioink thin film coated on top of a laser-transparent carrier, mostly a glass slide, which is sputtered with an energy-absorbing layer, for instance, gold [48]. A pulsed laser beam is focused on a spot of the absorbing layer, which rapidly heats up, evaporates locally, and generates a high gas pressure. The gas pressure in turn expands quickly and ejects a small fraction of the underlying bioink thin film towards the collector slide. For instance, Nd:YAG-laser with a wavelength of 1064 nm [48] or argon fluoride laser (ArF) with a wavelength of 193 nm can be applied as an energy source [49].

Laser-assisted bioprinting enables precise deposition of pico- to nanoliter drops and highly accurate features sizes ranging from 20 to 140μm [28, 50]. Besides, the LIFT mechanism offers nozzle-free material transfer, which strongly broadens the range of printable material viscosities (1–8279 mPa.s) compared to other drop-based printing methods [28, 41]. However, the process control is complex and its quality with respect to cell viability and drop size strongly depends on the applied material and laser beam settings, e.g. viscosity, energy absorption layer, wavelength, pulse duration, repetition rate, optical lenses, and the corresponding laser pulse energy. For these reasons, the achievable post-printing cell viability strongly varies with the applied settings and ranges from 64% to more than 95% [41, 51]. Also, most experimental laser-bioprinting setups require special security precautions to protect users from accidental laser light exposure. In conclusion, laser-assisted bioprinting requires intensive training and process knowhow to fully exploit its potential.

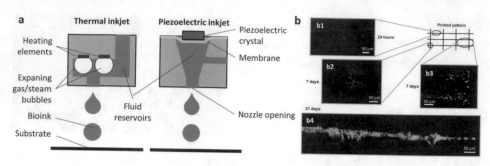

Fig. 9.7 Schematic illustration of the thermal (left) and piezoelectric (right) inkjet-based 3D-bioprinting process (**a**), adapted from Tseng and coworkers [52] as well as Laser and Santiago [53], and reproduced from the Biofabrication and 3D-bioprinting lecture notes with permission of Prof. Blaeser, Institute for BioMedical Printing Technology, Technical University Darmstadt. Cui and Boland applied inkjet-based bioprinting technology for the fabrication of microvasculature mimicking endothelial cell-laden fibrin channels (**b**) [54], reproduced and modified with permission from Elsevier

9.3.1.4 Inkjet Bioprinting

Inkjet bioprinting is a contactless printing method that uses the concept of conventional desktop printers to produce single drops of cell-loaded hydrogels. One of the first inkjet-bioprinters was developed in 2003 by Wilson and Boland [5]. According to the droplet ejection mechanism, thermal and piezoelectric inkjet bioprinters are distinguished (Fig. 9.7). In thermal inkjet bioprinting, small heating elements are used to create a rapidly expanding gas bubble that accelerates a fraction of bioink and ejects it through the nozzle [28, 55–57]. During the first 2–4 microseconds, peak temperatures of up to 300 °C and a pressure of up to 100 bar can be reached [56]. In piezoelectric inkjet systems an electro-mechanically driven diaphragm is deformed to print drops of bioink [58–61].

In inkjet bioprinters, the fluid is contained in a reservoir connected to a constantly opened nozzle. The surface tension of the fluid prevents nozzle dripping. Due to the open nozzle architecture, inkjet systems are mostly operated at ambient or vacuum pressure. This circumstance as well as the comparably narrow channels that connect the reservoir and nozzle hampers the flow of viscous fluids in a reasonable time. Most inkjet bioprinters are therefore limited to the processing of bioinks with viscosities ranging from 1.5 to 12 mPa.s [28, 41]. Within this process window, inkjet technology enables precise printing of pico- to nanoliter drops with feature sizes ranging from 30 to 300μm [28, 41, 50]. Post-printing cell survival rates ranging from 69 to 92% were reported [28, 41, 62].

9.3.1.5 Microvalve-Based Bioprinting

Microvalve-based bioprinting methods are often confused with inkjet bioprinting but differ in the underlying droplet formation mechanism and the printer head design. The printheads usually consist of two components: (1) a pneumatically pressurized fluid reservoir and (2) a switchable microvalve (Fig. 9.8). By briefly opening and closing the valve for a few

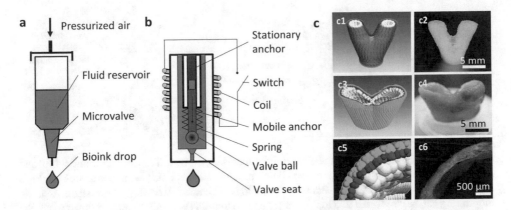

Fig. 9.8 Schematic illustration of a microvalve-based 3D-bioprinting cartridge (**a**) and the components of an electro-magnetic microvalve (**b**), adapted from Gyger et al. [63] and reproduced from the Biofabrication and 3D-bioprinting lecture notes with permission of Prof. Blaeser, Institute for BioMedical Printing Technology, Technical University Darmstadt. Blaeser and coworkers applied microvalve-based bioprinting technology for the fabrication of multimaterial branching vascular (C1, C2) and heart ventricle mimicking (C3–C6) alginate structures (**c**) [19] reproduced and modified with permission from John Wiley and Sons

microseconds, individual droplets can be generated and ejected from the nozzle. Electro-magnetically controlled microvalves or solenoid valves are at the heart of this process. The valve consists of a valve seat and a valve ball attached to a mobile anchor. By applying an electrical voltage to a coil surrounding the valve, it opens. The coil generates a magnetic field, which pulls the mobile anchor towards the stationary one against the force of a spring. When the power supply is interrupted, the magnetic field is reduced and the compression spring presses anchor and ball back, closing the valve.

In contrast to inkjet printer heads, which are constantly open, the nozzle can be controllably opened and closed in microvalve-based bioprinting. For this reason, the fluid reservoir can be operated at overpressures in the range of 0.5–3.0 bar without nozzle dripping [19, 64]. As a result, a broader spectrum of bioink viscosities (< 1500 mPa.s) can be printed [28]. Also, the nozzle size (100–600µm) is usually larger compared to inkjet printing (22–120µm) [65]. The achievable bioprinting resolution and cell viability rates depend on the applied bioink, nozzle size, gating time of the valve, and applied printing pressure. Drop volumes of 2–720 nL and feature sizes ranging from 100 to 1000µm can be generated with high precision [28, 50, 65]. The shear stress at the nozzle and, herewith, the post-printing cell viability can be controlled by fine-tuning the before mentioned influence factors. Correctly adjusted cell damage can be prevented and viabilities ranging from 71 to 99% can be achieved [28].

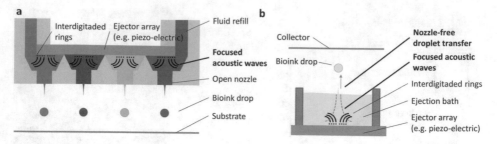

Fig. 9.9 Schematic illustration of acoustic bioprinting, adapted from Tasoglu and Demirci [50] and reproduced from the Biofabrication and 3D-bioprinting lecture notes with permission of Prof. Blaeser, Institute for BioMedical Printing Technology, Technical University Darmstadt. Acoustic bioprinting can be conducted in a top-down (**a**) and bottom-up (**b**) setup. In the top-down setup, the fabrication process is similar to inkjet bioprinting. The bottom-up approach enables nozzle-free bioprinting

9.3.1.6 Acoustic Bioprinting

Similar to the described LIFT process (Sect. 9.3.1.3), acoustic bioprinting is a contactless, nozzle-free bioprinting method. Acoustic waves are applied to transport small volumes of bioink from a fluid bath towards an opposing collector [66]. Piezoelectric elements are used to excite interdigitated ring elements that generate circular, surface acoustic waves [65]. These waves are focused on a defined point at the liquid interface. When the transported energy surpasses the surface tension of the liquid precursor, individual droplets are formed and ejected. The method was reported to be used along and also against ("upside-down") gravity for contactless transfer of liquids (Fig. 9.9) [67]. Nanoliter drops (45–300 nL) and feature sizes of approximately 250μm could be achieved with a post-printing cell viability of 90% [66, 68]. So far, comprehensive data is missing to narrow down the viscosity range of fluids and bioinks printable with this method.

9.3.1.7 Spheroid and Tissue Strand Based Bioprinting

In contrast to the aforementioned technologies, spheroid-based bioprinting, sometimes referred to as scaffold-free bioprinting, describes methods to print cell clusters and spheroids instead of cells encapsulated in hydrogels. Multicellular spheroids are clusters of 300–10,000 single cells with a size of 200–500μm [69–72]. They can be prepared in different ways, e.g. by the hanging drop method, in microfabricated molds, or rotary bioreactors [73]. Spheroids are mostly used as 3D cell culture models for high-throughput-screening of drugs or toxins [72]. However, they also resemble the native building blocks of tissues, which can be conveniently explored for the fabrication of three-dimensional engineered tissues. Different methods for the gentle deposition of spheroids have emerged in the past years. For instance, a modified extrusion method can be applied to deposit cell clusters. Here, the spheroids are collected, lined-up in a micropipette, and gently pushed out of it one-by-one [74]. Additional biomaterial or hydrogel scaffolds can be printed in parallel to fix the spheroid position [75]. Spheroids can also be printed in a pick-and-place

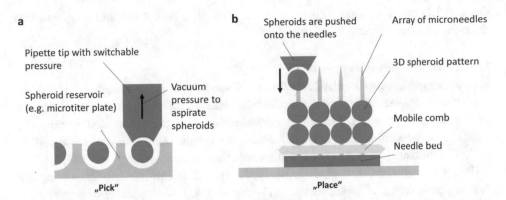

Fig. 9.10 Schematic illustration of a spheroid-based bioprinting method (the "Kenzan Method"), adapted from On et al. [78] and reproduced from the Biofabrication and 3D-bioprinting lecture notes with permission of Prof. Blaeser, Institute for BioMedical Printing Technology, Technical University Darmstadt. In a first step, pre-cultured multicellular spheroids are aspirated one-by-one from a microtiter plate (**a**). Next, the spheroids are placed in a predefined pattern onto an array of microneedles (**b**)

fashion (Fig. 9.10). In this case, individual spheroids are collected with a pipette, transported to an array of microneedles, and pinned on one of these [76]. By repetitive pinning of multiple spheroids on different spots of the needle array, cylindric 3D-geometries can reliably be fabricated. Finally, instead of spheroids another scaffold-free method applies several millimeter long tissue strands that can be produced and extruded to generate self-assembling tissue units [77].

> **Questions**
> 6. Please explain according to what criteria different bioprinting methods can be classified and give an example for each?
> 7. What quality criteria can be applied to compare different printing methods and which factors affect printing quality?

9.3.2 3D-Bioprinting Strategies

Due to their high water content and similarity to the natural extracellular matrix, hydrogels are a particularly suitable material for 3D cell culture and bioprinting. However, compared to thermoplastic polymers used in conventional rapid prototyping, such as PLA or PCL, hydrogels are mechanically weak and have a low Young's moduli. Besides, polymerization and gelation mechanisms of hydrogels are often time-consuming. Both can limit shape fidelity and inhibit the rapid fabrication of complex 3D-geometries. While the previous chapter focused on general mechanisms for dispensing, extruding, and printing bioinks

(Sect. 9.3.1), the following sections elucidate strategies, how they can be brought into three-dimensional shapes.

9.3.2.1 Bioink Reinforcement

One strategy to improve mechanical behavior, shape fidelity, gelling kinetics, and production speed is the mechanical reinforcement of bioinks. In literature, various approaches of mechanical reinforcement, such as dual crosslinking, the introduction of supramolecular bonds, interpenetrating gel networks, or nanocomposite reinforcement are described [79]. For instance, in dual crosslinking synthetic functional groups are added to the hydrogel to enable covalent binding, e.g. methacrylated gelatin [80]. Another example is the supplementation of bioinks with nanocomposite, e.g. fibers, particles, or molecules [81–84]. Blending ECM derived materials, such as collagen, with comparably high mechanical resilient and fast gelling hydrogels, e.g. polysaccharides like alginate or agarose, is a further popular approach [18, 85, 86]. Precisely adjusted, complex 3D-geometries with high aspect ratios have been generated. However, it should be noted that changes in the mechanical behavior, microstructure, and rheology of bioinks inevitably affect cell response.

9.3.2.2 Hybrid Bioprinting

Hybrid bioprinting, also known as co-printing or thermoplastic reinforcement, describes the parallel fabrication of bioinks and rigid biopolymers (e.g. PCL) [79]. Using the biopolymer, a stable and macro-porous framework is generated, which is infiltrated with the parallel printed bioink. In this way, a mechanically resilient 3D structure can be built up, while a high level of biofunctionality is achieved. The rigid biopolymer is usually applied using microextrusion, fused fiber fabrication, electrospinning, or melt-electro writing technology [20, 87–89]. In 2016, Anthony Atala and his research group used different hybrid bioprinting approaches to produce macroscopic, vascularized hydrogel structures mimicking bone, cartilage, and muscle tissue that ultimately could be implanted successfully in animal models [90].

9.3.2.3 Bioprinting with Solid or Liquid Support

Support materials enable fabrication of complex, high aspect ratio bioink structures. In contrast to the previously described hybrid bioprinting methods, the support material does not remain in the printed structure, but it is thermally, chemically, or mechanically detached from the bioprinted object. Bioprinting with solid and liquid support materials is distinguished. Thermally reversible hydrogels such as agarose, gelatin, or poloxamers are often used as solid support materials [19, 39, 75, 91]. They are either used as outer boundary layer for external stabilization during printing or as fugitive core to produce hollow structures and channels. In connection with spheroid-based, scaffold-free printing processes (Sect. 9.3.1.7), a further strategy has recently been developed, which to a certain extent can also be counted to printing methods using solid support. Instead of hydrogels or polymers, an array of metal needles is used to impale and keep in shape layer-by-layer applied spheroids [76].

Submerged bioprinting describes the application of liquids to support the three-dimensional fabrication of bioink structures. In this case, the printed object is placed on a vertically moveable platform, which is layer-wise lowered into a liquid-filled container. This approach is frequently applied for the fabrication of hydrogels that undergo physical or chemical crosslinking in the presence of a gelation agent. For example, alginate hydrogel can be printed in a bath of calcium chloride, which triggers its gelation [19, 92] or vice versa [93]. High-density, water-immiscible fluids, e.g. perfluorocarbons, are a choice to support fabrication of, for instance, thermo-gelling bioinks. Compared to other liquids, they possess a high affinity to bind respiratory gases, which is beneficial to provide sufficient oxygen to cells in pro-longed fabrication times [94, 95]. Hydrogel structures with large aspect ratios and overhanging branches can be fabricated using this method [17, 96]. Finally, thermo-reversible hydrogel precursor solutions, such as swollen gelatin microparticles, can be used as pseudoplastic support [97]. In recent studies, the described method was successfully applied to fabricate complex, anatomical structures from comparably soft, protein-based bioinks, such as collagen [98]. Depending on the bioink to be processed, the described liquid and solid support strategies can be applied individually or combined [19].

9.3.2.4 Core–Shell Modeling

Core–shell modeling is an advanced microextrusion method that differs in the composition of the printed material. Instead of a single material, two or more components are co-extruded to generate, radially stratified core–shell strands. The concept can be used to encapsulate softer, low-concentrated bioinks in a denser, more rigid shell. This way, the mechanically resilient shell material enables fast fabrication of robust 3D-geometries, while the core provides a biofunctional environment to stimulate cell growth [44]. Alternatively, the method can be used to fabricate vascular channel resembling hollow strands by co-extrusion of a solid or liquid support material together with a bioink shell [81, 99, 100]. The same method can also be applied to administer a crosslinking agent either in the core or the shell of a printed strand [43].

Questions
8. Which bioprinting strategies can be distinguished?
9. What is the reasoning for applying these strategies?

9.4   Maturation and Application of Bioprinted Tissue

The first decade of bioprinting research was dedicated to the detailed study, development, and advancement of bioprinting technologies, which have been broadly described in the previous chapters. A couple of years ago, scientists have started to take the first steps towards pre-clinical and industrial translation of this inspiring technology. Concerning their application in biomedicine, bioprinted tissue substitutes that can be implanted in living organisms and in vitro tissue models for drug and toxicity screening are distinguished (Fig. 9.11). Also, bioprinting technology can be applied to the production of food and consumables, such as meat or leather.

Following the bioprinting process, the fabricated units must be cultivated to form functional tissues. Tissue formation under static and dynamic condition are herein distinguished. For dynamic cultivation, fluid dynamic stress, mechanical load, or electrical impulses are applied to stimulate and condition tissue growth. Depending on the targeted tissue type and application, cultivation in macroscopic bioreactors and microfluidic devices is distinguished. The following paragraphs will highlight different target tissues and cultivation methods in both the fabrication of tissue implants and in vitro models.

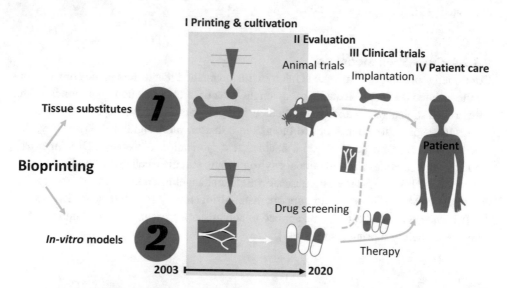

Fig. 9.11 Illustration of bioprinting fields of research and steps towards clinical application, adapted from Blaeser et al. [28] and reproduced from the Biofabrication and 3D-bioprinting lecture notes with permission of Prof. Blaeser, Institute for BioMedical Printing Technology, Technical University Darmstadt. Two biomedical applications are distinguished. Tissue substitutes are intended to be used for implantation to substitute, repair, or maintain a damaged tissue or organ. In vitro tissue models can be applied for drug and toxicity testing as well as disease modeling

9.4.1   Bioprinted Tissue Substitutes

This subsection presents bioprinted tissue structures that are intended for use as implants to restore, substitute, or maintain the function of damaged tissues or organs [1, 101]. For this purpose, a variety of tissues have recently been fabricated using 3D bioprinting technologies. Among those, tissues that address the locomotor apparatus, such as cartilage and bone tissue predominate [28, 90, 102–106]. Besides these, efforts towards the biofabrication of vasculature [75], heart muscle [98], tendon [107], skin [108], dental [23], neuronal [109], as well as tracheal [110] and pancreatic tissues [111, 112] have been made (Fig. 9.12, Table 9.1) [28, 121].

9.4.1.1   Design Considerations and Pre-Conditioning

The goal of creating tissue substitutes is their implantation into a host organism, such as an animal model or patient. Besides biological functionality, tissue substitutes need to exhibit minimal mechanical requirements to withstand the implantation surgery and naturally occurring load at the implantation site. To foster both, biological functionality and mechanical resilience, tissue implants are mostly pre-conditioned in vitro before implantation [122, 123]. Pre-conditioning under static and dynamic culture conditions can be herein distinguished. In specific cases, tissue precursors that already match the required strength after printing or those that are mechanically reinforced can be directly implanted and further conditioned in vivo using the host's body as a natural bioreactor [124]. Load bearing implants, such as cartilage, bone, or muscle, are commonly pre-conditioned in

Bioprinted tissue substitutes

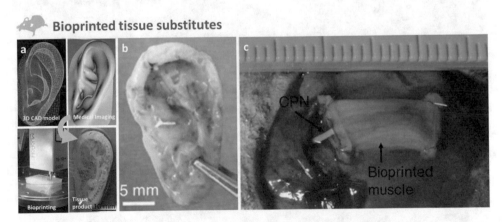

Fig. 9.12 Exemplary presentation of bioprinted tissue substitutes for implantation reproduced from Blaeser and coworkers [28] as well as Kang and coworkers with permission of Elsevier and Springer Nature. Based on CAD or medical imaging data tissue substitutes were fabricated using a hybrid 3D-bioprinting approach (**a**) [90]. The printed objects were implanted in animal models for several weeks to mature towards ear shaped cartilage (**b**), and muscle (**c**) tissue [90]

Table 9.1 Bioprinting tissue substitutes, adapted from Blaeser et al. [28] and reproduced from the Biofabrication and 3D-bioprinting lecture notes with permission of Prof. Blaeser, Institute for BioMedical Printing Technology, Technical University Darmstadt with permission from Elsevier. The table summarizes recently published work that reports on the fabrication of tissue substitutes using bioprinting technology

Examples		Material	Cell types	In vivo assessment	Culture
Bone	[102]	RGD-Alginate, PEGMA, GelMA, PCL	Murine MSC	Bone volume/area after implantation (μCT imaging)	12 weeks (mouse)
	[18]	Agarose, Collagen	hMSC	n. a.	3 weeks (in vitro)
	[113]	PLA, GelMA	hMSC, HUVECs	n. a.	4 weeks (in vitro)
	[114]	PEGDA, GelMA,	hMSC	n. a.	3 weeks (in vitro)
	[90]	PCL, Gelatin, HA, Fibrinogen, Glycerol	Human AFSCs	Bone/osteoid formation, blood vessel formation	5 months (rat)
Cartilage	[77]	Alginate	Chondrocytes	n. a.	3 week (in vitro)
	[115]	PEG, polyHPMA-lac, CSMA, HAMA	Chondrocytes	n. a.	6 weeks (in vitro)
	[116]	GelMA, CS-AEMA, HAMA, Alginate	hMSC	n. a.	3 weeks (in vitro)
	[117]	PEGDA	hMSC	n. a.	3 weeks (mouse)
	[90]	PCL, Gelatin, HA, Fibrinogen, Glycerol	Rabbit ear chondrocytes	Ear cartilage reconstruction, blood vessel formation	2 months (mouse)
Muscle tissue	[107]	TPU, PCL, Gelatin, Fibrinogen, HA	Myoblasts, fibroblasts	n. a.	1 week (in vitro)
	[90]	PCL, Gelatin, HA, Fibrinogen, Glycerol	Myoblasts	Muscle function (electromyography), innervating capability	2 weeks (rat)

(continued)

Table 9.1 (continued)

Examples		Material	Cell types	In vivo assessment	Culture
Neural tissue	[118]	Polyurethane dispersions	Murine NSCs	Rescue function after traumatic brain injury (locomotion, survival)	10 days (zebra fish)
Osteochondral	[119]	PCL, Atelocollagen, HA	hTMSC	Evaluation of neocartilage by ICRS scoring system	8 weeks (rabbit)
Pancreas	[112]	PCL, Alginate	HUVECs, Islets of Langerhans,	Glucose induced insulin secretion test	4 days (CAM assay)
Sweat glands	[120]	Gelatin, Alginate	Murine EPCs	Functional restoration of sweat glands in mice	2 weeks (mouse)

macroscopic perfusion bioreactors, which additionally exercise mechanical stress [122, 125, 126]. Implants of the cardio-vascular system, e.g. blood vessels or heart valves, are conditioned under dynamic and sometimes pulsatile flow [75, 127, 128]. Epithelialized tissues like skin are mostly cultivated with an air-liquid interface [108, 129].

9.4.1.2 In Vivo Evaluation

Recent studies have shown that many bioprinted tissue precursors are studied in vivo either directly after printing or following a static culture of several days to weeks [90]. The potential to structurally reinforce engineered tissues using different bioprinting strategies (Sect. 9.3.2) favors this trend. For example, it is reported that most bioprinted tissue implants are fabricated using solid [76] and liquid support materials (Sect. 9.3.2.3) [98] or hybrid bioprinting methods (Sect. 9.3.2.2) [28, 130]. On average, so far printed constructs comprise three different types of material, such as hydrogels and thermoplastic polymers, and one cell type [28]. It is reported that about one half of the conducted studies assessed the performance of the bioprinted tissue precursors for 10 days to 5 months in vivo. So far, mice, rats, zebrafish, and rabbits are preferred animal models for implantation [28]. Previous in vivo implantations have been a good indicator for the clinical relevance of bioprinted tissue substitutes. In a recent study, researchers observed mature bone and osteoid formation as well as blood vessel sprouting in bone tissue 5 months post-implantation [28, 90]. Successful reconstruction of ear cartilage, as well as the fabrication of functional, innervated muscle tissue was also reported [28, 90].

9.4.2   Bioprinted In Vitro Models

Besides their suitability as implantable substitutes, bioprinted tissue units can be used as predictive in vitro models for drug screening and toxicity testing. Several healthy, as well as diseased tissue models, have recently been fabricated. These include models for alveolar, bone, cardiac, gut, hepatic, neural, renal, skin tissue, and vascular systems [28, 131–133]. Also, diseased tissues, such as brain tumor, breast cancer metastasis, or thrombosis, have been replicated (Fig. 9.13, Table 9.2) [28].

9.4.2.1 Design Considerations

Due to their prospective application in pharmacological and toxicological research, bioprinted in vitro models are generally subject to other physical, biological, and economical requirements and design considerations compared to the previously described tissue implants. Since they are not intended for implantation, they do not have to exhibit the same

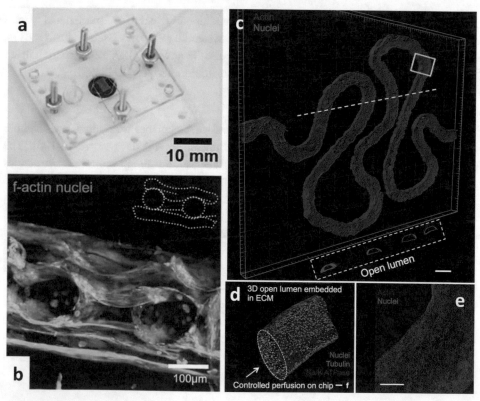

Fig. 9.13 Exemplary presentation of bioprinted in vitro tissue models for drug- and toxicity testing reproduced from Blaeser and coworkers [28], Zhang et al. [134], and Homan and coworkers [135] with permission of Elsevier and Springer Nature. A cardiac tissue-mimicking model was printed into a microfluidic chamber (**a, b**) [134]. In another study the renal proximal tubule was reproduced using 3D-bioprinting technology (**c–e**) [135]

Table 9.2 Bioprinting in vitro models, adapted from Blaeser et al. [28] and reproduced from the Biofabrication and 3D-bioprinting lecture notes with permission of Prof. Blaeser, Institute for BioMedical Printing Technology, Technical University Darmstadt with permission from Elsevier. The table summarizes recently published work that reports on the bioprinting of in vitro tissue models for drug and toxicity screening

Tissue	Cited work	Material	Cell types	Clinical readout	Culture
Alveolar tissue	[136]	Matrigel	Endothelial and epithelial cells	n. a.	5 days
Bone tissue	[113]	PLA, Gelatin, Polylysine	hMSC, HUVECs	n. a.	28 days
Brain tumor	[137]	Gelatin, Fibrin Alginate	Glioma stem cells, Human glioma cells	Tumor sensitivity to the chemotherapeutic Temozolomide	21 days
Breast cancer	[138]	GelMA + nHA	BrCa cells, fetal osteoblasts, hMSC	n. a.	14 days
Cardiac tissue	[139]	GelMA, Alginate	HUVECs, cardiomyocytes	Influence of Doxorubicin on beating rate and vWF Level	15 days
	[140]	Dextran, TPU, ABS, PDMS, Ag:PA, PLA	NRVMs, hiPS-CMs	Influence of Verapamil and Isoproterenol on the contraction force (inotropic response)	28 days
Hepatic tissue	[141]	HA, Gelatine, PEG	Hepatocytes, stellate cells, Kupffer cells	Albumin and urea production	20 days
	[142]	GelMA	Hepatic progenitor cells, ASCs, HUVECs	n. a.	20 days
	[143]	PCL, gelatin, collagen	HepG2, HUVEC	Albumin and urea production	6 days
	[144]	NovoGel® 2.0	Human hepatocytes, stellate cells, HUVECs	Influence of Trovafloxacin and Levofloxacin on LDA, albumin, and ATP	28 days
Neural tissue	[145]	Alginate, Agarose, CMC	hNSCs	Bicuculline-induced increased calcium response	10 days
Renal tissue	[135]	Gelatin, Fibrin Pluronic	PTE cells, fibroblasts, immune cells	Influence of Cyclosporine A on epithelial viability and barrier function	42 days

(continued)

Table 9.2 (continued)

Tissue	Cited work	Material	Cell types	Clinical readout	Culture
Skin tissue	[146]	Collagen	Dermal fibroblasts	Transdermal penetration ability of nanoparticles	7 days
Thrombosis	[147]	GelMA	Fibroblasts, HUVECs	Thrombolysis by tPA flow	14 days

mechanical resilience required for surgery and do not have to comply with mechanical load experienced after implantation. The biological challenges are different, too. For instance, tissue models are frequently generated from genetically modified cells, e.g. induced-pluripotent stem cells, which are rather unlikely applied for transplantation [78]. However, expectations regarding their biological response are rather high. To deliver relevant testing data, they must recapitulate specific physiological functions of a tissue or organ unit as close as possible. The composition of fabricated tissue models that have been reported reflects this. In contrast to tissue implants, so far fabricated tissue models comprised on average two different materials and two different cell types, indicating that biological complexity outweighs mechanical support considerations [28]. Moreover, economic aspects should also be taken into account when fabricating bioprinted tissue models [148]. For drug screening and toxicity tests, medium and high-throughput methods are applied, which require rapid, large-scale production of models using as few resources as possible [72].

For the above-mentioned reasons, the fabrication, post-printing cultivation, and conditioning methods of bioprinted tissue models differ from those described for tissue substitutes (Sect. 9.4.1). Most tissue models are either printed into well-plates or Petri-dishes for static cultivation or are printed into microfluidic chambers or chips, also referred to as organ-on-a-chip, wherein they can be dynamically cultured and evaluated [28]. The models can be fabricated in-situ or transferred into these chambers after printing. The parallel freeform fabrication of tissue model and chip has also been reported [143]. In dynamic cultivation, the cell culture medium is supplied either extrinsically by channels integrated into the chamber or intrinsically by vascularized structures of the model connected to the inlet and outlet of it.

9.4.2.2 Pre-Cultivation and Screening Studies

Before application as a drug and toxicity screening platform, the models are pre-cultivated for one to several weeks to allow maturation towards a physiologically responding tissue of interest. The expression of specific morphological, biological, and physical markers is often used as an indicator to determine the completion of pre-cultivation and to set the starting point of the screening study. These include, for example, confluent lining of surfaces and channels with endothelial and epithelial cells; tissue-specific gene expression and metabolic activity, such as secretion of albumin and urea in liver models; and the exertion of mechanical forces, like contraction of heart muscle cells in myocardial models. A recent study has shown the development of a projection stereolithography printed model of the liver lobe, which exhibited liver-specific gene expression levels, increased metabolic product secretion, and enhanced cytochrome P450 induction [30].

Finally, specific markers are used as a starting point (controls) to investigate the effect of a certain drug or toxin. Even though the development of bioprinted tissue models is still in an early stage, initial studies already obtained clinically relevant readouts [28]. A successful example is the sensitivity testing of tumor cells to chemotherapeutics, such as Temozolomide, which was, for instance, tested in a brain tumor model [137]. A bioprinted renal proximal tube was used to study the barrier function of renal epithelium and the effect of nephrotoxin exposure, e.g. by Cyclosporine A administration [135]. Furthermore, the effect of Doxorubicin, Verapamil, and Isoproterenol on the beating rate of cardiac tissue or the toxicity of Trovafloxacin and Levofloxacin on bioprinted liver tissue was investigated [140, 144, 149].

Questions

10. How do design considerations and constrictions differ for bioprinted implants and tissue models?
11. Please list five examples of so far fabricated tissue implants and models as well as their respective in vivo assessment criteria or readouts, respectively.

Answers

1. Tissue engineering combines biological, medical, and engineering approaches to recreate living structures, which can be used to substitute, restore, or maintain the function of damaged tissues and organs.
2. Automation and up-scaling of manual tissue engineering procedures were the initial reasons to apply bioprinting technology. Besides automation, its application was further motivated to benefit from additional advantages promised by 3D-printing technology, such as spatially controlled, reproducible deposition of different bioinks as well as the fabrication of complex and patient-individualized 3D-geometries.
3. Transformation of medical imaging data into 3D-printable commands requires segmentation, tessellation (e.g. STL-file format), slicing, and machine code translation (e.g. G-code).
4. Bioinks are defined as a mixture of living cells and a printable biomatrix, such as a hydrogel. Further supplements, such as growths factors, peptides, fibers, or particles might be added.

(continued)

5. Alginate, agarose, and collagen type I.
6. Depending on the scale of bioink transfer, three different classes of bioprinting methods are distinguished, layer-wise (e.g. projection stereolithography), line-wise (microextrusion), and drop-wise printing (e.g. microvalve-based bioprinting).
7. The quality of bioprinting methods is commonly described by the printing resolution (e.g. minimally achievable feature size, drop volume, or drop size), the viscosity range of printable bioink, and the post-printing cell viability. Further criteria that are less frequently compared are the printing speed, drop frequency, and throughput. The described criteria do not only depend on the applied printing method but are strongly influenced by the bioink rheology (e.g. viscosity, shear-thinning behavior, and surface tension), the cell type, the applied printing settings (e.g. pressure, nozzle size, temperature), and the experience of the user.
8. The following bioprinting strategies can be distinguished: bioink reinforcement, hybrid bioprinting, bioprinting with solid or liquid support.
9. Hydrogels are a particularly suitable material for 3D cell culture and bioprinting but lack mechanical resilience and shape fidelity. To still generate complex 3D-structures and geometries specific bioprinting strategies can be applied.
10. Tissue substitutes are meant to be implanted and therefore have to exhibit not only biological functionality and compatibility, but also sufficient mechanical resilience to withstand surgery and post-implantation load. In contrast, design considerations for bioprinted in vitro models are less focused on mechanics and more driven by biological complexity, and economical production methods.
11. See Tables 9.1 and 9.2.

Take-Home Message
- Bioprinting is a versatile additive manufacturing technology for the fabrication of complex tissue units.
- In the field of medicine, bioprinting technologies are used for the fabrication of implants and tissue models.
- Automation and scalability, spatially controlled and reproducible deposition of different bioinks, as well as the fabrication of individualized 3D-geometries and implants, are the main advantages of 3D-bioprinting.

(continued)

- A one-fits-all bioprinting method or strategy does not exist. According to the desired bioink, cell type, and application, each method and strategy has its benefits and disadvantages.
- The bioprinting method and strategy as well as the design considerations and post-printing cultivation methods need to be equally weighted for obtaining successful outcomes.

References

1. Langer R, Vacanti JP. Science. 1993;260:920.
2. Atala A, Bauer SB, Soker S, Yoo JJ, Retik AB. Lancet. 2006;367:1241.
3. Weber M, Gonzalez de Torre I, Moreira R, Frese J, Oedekoven C, Alonso M, Rodriguez Cabello CJ, Jockenhoevel S, Mela P. Tissue Eng Part C Methods. 2015;21:832.
4. Nieponice A, Soletti L, Guan J, Deasy BM, Huard J, Wagner WR, Vorp DA. Biomaterials. 2008;29:825.
5. Wilson WC, Boland T. Anat Rec A Discov Mol Cell Evol Biol. 2003;272:491.
6. Pandey PM, Reddy NV, Dhande SG. Int J Mach Tools Manuf. 2003;43:61.
7. Ding D, Pan Z, Cuiuri D, Li H, van Duin S. Advanced design for additive manufacturing: 3d slicing and 2d path planning. In: New trends 3D print. Rijeka: InTech; 2016.
8. Szilvśi-Nagy M, Mátyási G. Math Comput Model. 2003;38:945.
9. Wu T, Cheung EHM. Int J Adv Manuf Technol. 2005;29:1143.
10. Eragubi M. Int J Innov Manag Technol. 2013;4:410.
11. Petitjean C, Dacher J-N. Med Image Anal. 2011;15:169.
12. Heimann T, Meinzer H-P. Med Image Anal. 2009;13:543.
13. Blaeser A, Heilshorn SC, Duarte Campos DF. Gels. 2019;5:29.
14. Zhang YS, Yue K, Aleman J, Mollazadeh-Moghaddam K, Bakht SM, Yang J, Jia W, Dell'Erba V, Assawes P, Shin SR, Dokmeci MR, Oklu R, Khademhosseini A. Ann Biomed Eng. 2017;45:148.
15. Groll J, Burdick JA, Cho DW, Derby B, Gelinsky M, Heilshorn SC, Jüngst T, Malda J, Mironov VA, Nakayama K, Ovsianikov A, Sun W, Takeuchi S, Yoo JJ, Woodfield TBF. Biofabrication. 2019;11:013001. https://doi.org/10.1088/1758-5090/aaec52.
16. Ashammakhi N, Ahadian S, Xu C, Montazerian H, Ko H, Nasiri R, Barros N, Khademhosseini A. Mater Today Bio. 2019;1:100008.
17. Duarte Campos DF, Blaeser A, Weber M, Jäkel J, Neuss S, Jahnen-Dechent W, Fischer H. Biofabrication. 2013;5:015003.
18. Duarte Campos DF, Blaeser A, Buellesbach K, Sen KS, Xun W, Tillmann W, Fischer H. Adv Healthc Mater. 2016;5:1336.
19. Blaeser A, Duarte Campos DF, Puster U, Richtering W, Stevens MM, Fischer H. Adv Healthc Mater. 2016;5:326.
20. Campos DFD, Philip MA, Gürzing S, Melcher C, Lin YY, Schöneberg J, Blaeser A, Theek B, Fischer H, Betsch M, Print 3D. Addit Manuf. 2019;6:63.
21. Duarte Campos DF, Blaeser A, Korsten A, Neuss S, Jäkel J, Vogt M, Fischer H. Tissue Eng Part A. 2015;21:740.

22. Duarte Campos DF, Rohde M, Ross M, Anvari P, Blaeser A, Vogt M, Panfil C, Yam GH, Mehta JS, Fischer H, Walter P, Fuest M, Biomed J. Mater Res Part A. 2019;107:1945.
23. Duarte Campos DF, Zhang S, Kreimendahl F, Köpf M, Fischer H, Vogt M, Blaeser A, Apel C, Esteves-Oliveira M. Connect Tissue Res. 2019;61:205.
24. Campos DFD, Marquez AB, O'seanain C, Fischer H, Blaeser A, Vogt M, Corallo D, Aveic S. Cancers (Basel). 2019;11:2.
25. Morris VB, Nimbalkar S, Younesi M, McClellan P, Akkus O. Ann Biomed Eng. 2017;45:286.
26. Hockaday LA, Kang KH, Colangelo NW, Cheung PYC, Duan B, Malone E, Wu J, Girardi LN, Bonassar LJ, Lipson H, Chu CC, Butcher JT. Biofabrication. 2012;4:035005.
27. Wang H, Heilshorn SC. Adv Mater. 2015;27:3717–36.
28. Blaeser A, Duarte Campos DF, Fischer H. Curr Opin Biomed Eng. 2017;2:58.
29. Melchels FPW, Feijen J, Grijpma DW. Biomaterials. 2010;31:6121.
30. Ma X, Qu X, Zhu W, Li Y-S, Yuan S, Zhang H, Liu J, Wang P, Lai CSE, Zanella F, Feng G-S, Sheikh F, Chien S, Chen S. Proc Natl Acad Sci. 2016;113:2206.
31. Sun C, Fang N, Wu DM, Zhang X. Sensors Actuators A Phys. 2005;121:113.
32. Fouassier J, Allonas X, Burget D. Prog Org Coat. 2003;47:16.
33. Fedorovich NE, Oudshoorn MH, van Geemen D, Hennink WE, Alblas J, Dhert WJA. Biomaterials. 2009;30:344.
34. Han D, Yang C, Fang NX, Lee H. Addit Manuf. 2019;27:606.
35. Pati F, Jang J, Lee JW, Cho DW. Extrusion bioprinting. In: Essentials of 3D biofabrication and translation. Cambridge: Academic Press; 2015. p. 123–52.
36. Ozbolat IT, Hospodiuk M. Biomaterials. 2016;76:321.
37. Cohen DL, Malone E, Lipson HOD, Bonassar LJ. Tissue Eng. 2006;12:1325.
38. Kolesky DB, Truby RL, Gladman AS, Busbee TA, Homan KA, Lewis JA. Adv Mater. 2014;26:1.
39. Kolesky DB, Homan KA, Skylar-scott MA, Lewis JA. PNAS. 2016;113:3179.
40. Chang CC, Boland ED, Williams SK, Hoying JB. J Biomed Mater Res B Appl Biomater. 2011;98:160.
41. Murphy SV, Atala A. Nat Biotechnol. 2014;32:773.
42. Markstedt K, Mantas A, Tournier I, Martínez Ávila H, Hägg D, Gatenholm P. Biomacromolecules. 2015;16:1489.
43. Colosi C, Shin SR, Manoharan V, Massa S, Costantini M, Barbetta A, Dokmeci MR, Dentini M, Khademhosseini A. Adv Mater. 2016;28:677.
44. Akkineni AR, Ahlfeld T, Lode A, Gelinsky M. Biofabrication. 2016;8:45001.
45. Serra P, Piqué A. Adv Mater Technol. 2019;4:1.
46. Gruene M, Unger C, Koch L, Deiwick A, Chichkov B. Biomed Eng Online. 2011;10:1.
47. Guillotin B, Souquet A, Catros S, Duocastella M, Pippenger B, Bellance S, Bareille R, Rémy M, Bordenave L, Amédée J, Guillemot F. Biomaterials. 2010;31:7250.
48. Koch L, Kuhn S, Sorg H, Gruene M, Schlie S, Gaebel R, Polchow B, Reimers K, Stoelting S, Ma N, Vogt PM, Steinhoff G, Chichkov B. Tissue Eng Part C Methods. 2010;16:847.
49. Schiele NR, Corr DT, Huang Y, Raof NA, Xie Y, Chrisey DB. Biofabrication. 2010;2:032001.
50. Tasoglu S, Demirci U. Trends Biotechnol. 2013;31:10.
51. Xiong R, Zhang Z, Chai W, Huang Y, Chrisey DB. Biofabrication. 2015;7:45011.
52. Tseng F, Kim C, Ho C. J Microelectromech Syst. 2002;11:427.
53. Laser DJ, Santiago JG. J Micromech Microeng. 2004;14:R35.
54. Cui X, Boland T. Biomaterials. 2009;30:6221.
55. Sen AK, Darabi J. J Micromech Microeng. 2007;17:1420.
56. Chen P-H, Chen W-C, Chang S-H. Int J Mech Sci. 1997;39:683.
57. Xu T, Jin J, Gregory C, Hickman JJJJ, Boland T. Biomaterials. 2005;26:93.

58. Kim JD, Choi JS, Kim BS, Chan Choi Y, Cho YW. Polymer (Guildf). 2010;51:2147.
59. Lorber B, Hsiao W-K, Hutchings IM, Martin KR. Biofabrication. 2014;6:015001.
60. Wijshoff H. Phys Rep. 2010;491:77.
61. De Maria C, Ferrari L, Montemurro F, Vozzi F, Guerrazzi I, Boland T, Vozzi G. Procedia Eng. 2015;110:98.
62. Matai I, Kaur G, Seyedsalehi A, McClinton A, Laurencin CT. Biomaterials. 2020;226:119536.
63. Fritz Gyger AG. Produktbroschüre Mikroventile SMLD. 2010. www.fluidics.ch. Accessed 04 Aug 2015.
64. Faulkner-jones A, Fyfe C, Cornelissen D, Gardner J, King J. Biofabrication. 2015;7:44102.
65. Gudapati H, Dey M, Ozbolat I. Biomaterials. 2016;102:20.
66. Demirci U, Montesano G. Lab Chip. 2007;7:1139.
67. Ellson R, Mutz M, Browning B, Leejr L, Miller M, Papen R. J Assoc Lab Autom. 2003;8:29.
68. Fang Y, Frampton JP, Raghavan S, Sabahi-Kaviani R, Luker G, Deng CX, Takayama S. Tissue Eng Part C Methods. 2012;18:647.
69. Timmins NE, Dietmair S, Nielsen LK. Angiogenesis. 2004;7:97.
70. Grimes DR, Kelly C, Bloch K, Partridge M. J R Soc Interf. 2014;11:20131124.
71. Bhise NS, Manoharan V, Massa S, Tamayol A, Ghaderi M. Biofabrication. 2016;8:14101.
72. Kunz-Schughart LA, Freyer JP, Hofstaedter F, Ebner R. J Biomol Screen. 2004;9:273.
73. Lin R-Z, Lin R-Z, Chang H-Y. Biotechnol J. 2008;3:1172.
74. Jakab K, Norotte C, Marga F, Murphy K, Vunjak-Novakovic G, Forgacs G. Biofabrication. 2010;2:022001.
75. Norotte C, Marga FS, Niklason LE, Forgacs G. Biomaterials. 2009;30:5910.
76. Arai K, Murata D, Verissimo AR, Mukae Y, Itoh M, Nakamura A, Morita S, Nakayama K. PLoS One. 2018;13:1.
77. Yu Y, Moncal KK, Li J, Peng W, Rivero I, Martin JA, Ozbolat IT. Sci Rep. 2016;6:28714.
78. Ong CS, Yesantharao P, Huang CY, Mattson G, Boktor J, Fukunishi T, Zhang H, Hibino N. Pediatr Res. 2018;83:223.
79. Chimene D, Kaunas R, Gaharwar AK. Adv Mater. 2019;1902026:1.
80. Chen YC, Lin RZ, Qi H, Yang Y, Bae H, Melero-Martin JM, Khademhosseini A. Adv Funct Mater. 2012;22:2027.
81. Blaeser A, Million N, Duarte Campos DF, Gamrad L, Köpf M, Rehbock C, Nachev M, Sures B, Barcikowski S, Fischer H. Nano Res. 2016;9:3407.
82. Mirahmadi F, Tafazzoli-Shadpour M, Shokrgozar MA, Bonakdar S. Mater Sci Eng C Mater Biol Appl. 2013;33:4786.
83. Kosik-Kozioł A, Costantini M, Bolek T, Szöke K, Barbetta A, Brinchmann J, Święszkowski W. Biofabrication. 2017;9:044105. https://doi.org/10.1088/1758-5090/aa90d7.
84. Kumar BYS, Isloor AM, Kumar GCM, Inamuddin, Asiri AM. Sci Rep. 2019;9:1.
85. Köpf M, Campos DFD, Blaeser A, Sen KS, Fischer H. Biofabrication. 2016;8:1.
86. Kreimendahl F, Köpf M, Thiebes AL, Duarte Campos DF, Blaeser A, Schmitz-Rode T, Apel C, Jockenhoevel S, Fischer H. Tissue Eng Part C Methods. 2017;23:604.
87. Shim J-H, Kim JY, Park M, Park J, Cho D-W. Biofabrication. 2011;3:034102.
88. Yoon Y, Kim CH, Lee JE, Yoon J, Lee NK, Kim TH, Park SH. Biofabrication. 2019;11:025015. https://doi.org/10.1088/1758-5090/ab08c2.
89. de Ruijter M, Ribeiro A, Dokter I, Castilho M, Malda J. Adv Healthc Mater. 2019;8:e1800418. https://doi.org/10.1002/adhm.201800418.
90. Kang H-W, Lee SJ, Ko IK, Kengla C, Yoo JJ, Atala A. Nat Biotechnol. 2016;34:312.
91. Wu W, DeConinck A, a Lewis J. Adv Mater. 2011;23:H178.
92. Tabriz AG, Hermida MA, Leslie NR, Shu W. Biofabrication. 2015;7:045012.

93. Boland T, Xu T, Damon B, Cui X. Biotechnol J. 2006;1:910.
94. Radisic M, Park H, Chen F, Salazar-Lazzaro JE, Wang Y, Dennis R, Langer R, Freed LE, Vunjak-Novakovic G. Tissue Eng. 2006;12:2077.
95. Lowe KC. Tissue Eng. 2003;9:389.
96. Blaeser A, Duarte Campos DF, Weber M, Neuss S, Theek B, Fischer H, Jahnen-Dechent W. Biores Open Access. 2013;2:374.
97. Hinton TJ, Jallerat Q, Palchesko RN, Park JH, Grodzicki MS, Shue HJ, Ramadan MH, Hudson AR, Feinberg AW. Sci Adv. 2015;1:e1500758. https://doi.org/10.1126/sciadv.1500758.
98. Lee A, Hudson AR, Shiwarski DJ, Tashman JW, Hinton TJ, Yerneni S, Bliley JM, Campbell PG, Feinberg AW. Science. 2019;365:482.
99. Blaeser A, Duarte Campos DF, Köpf M, Weber M, Fischer H. RSC Adv. 2014;4:46460.
100. Jia W, Gungor-Ozkerim PS, Zhang YS, Yue K, Zhu K, Liu W, Pi Q, Byambaa B, Dokmeci MR, Shin SR, Khademhosseini A. Biomaterials. 2016;106:58.
101. Vacanti CA, Vacanti JP. Orthop Clin North Am. 2000;31:351.
102. Daly AC, Cunniffe GM, Sathy BN, Jeon O, Alsberg E, Kelly DJ. Adv Healthc Mater. 2016;5:2353.
103. Daly AC, Freeman FE, Gonzalez-Fernandez T, Critchley SE, Nulty J, Kelly DJ. Adv Healthc Mater. 2017;6:1.
104. Singh YP, Bandyopadhyay A, Mandal BB. ACS Appl Mater Interfaces. 2019;11:33684.
105. Rathan S, Dejob L, Schipani R, Haffner B, Möbius ME, Kelly DJ. Adv Healthc Mater. 2019;8:1.
106. Piard C, Baker H, Kamalitdinov T, Fisher J. Biofabrication. 2019;11:025013. https://doi.org/10.1088/1758-5090/ab078a.
107. Merceron TK, Burt M, Seol Y-J, Kang H-W, Lee SJ, Yoo JJ, Atala A. Biofabrication. 2015;7:035003.
108. Michael S, Sorg H, Peck C-T, Koch L, Deiwick A, Chichkov B, Vogt PM, Reimers K. PLoS One. 2013;8:e57741.
109. Yurie H, Ikeguchi R, Aoyama T, Kaizawa Y, Tajino J, Ito A, Ohta S, Oda H, Takeuchi H, Akieda S, Tsuji M, Nakayama K, Matsuda S. PLoS One. 2017;12:1.
110. Ke D, Yi H, Est-Witte S, George S, Kengla C, Atala A. Biofabrication. 2019;12:015022. https://doi.org/10.1088/1758-5090/ab5354.
111. Marchioli G, van Gurp L, van Krieken PP, Stamatialis D, Engelse M, van Blitterswijk CA, Karperien MBJ, de Koning E, Alblas J, Moroni L, van Apeldoorn AA. Biofabrication. 2015;7:025009.
112. Marchioli G, Di Luca A, De Koning E, Engelse M, Van Blitterswijk CA, Karperien M, Van Apeldoorn AA. Adv Healthc Mater. 2016;5:1606.
113. Cui H, Zhu W, Nowicki M, Zhou X, Khademhosseini A, Zhang LG. Adv Healthc Mater. 2016;5:395.
114. Yang F-Y, Yu M-X, Zhou Q, Chen W-L, Gao P, Huang Z. Cell Transplant. 2012;21:1805.
115. Abbadessa A, Mouser VHM, Blokzijl MM, Gawlitta D, Dhert WJA, Hennink WE, Malda J, Vermonden T. Biomacromolecules. 2016;17:2137.
116. Costantini M, Idaszek J, Szöke K, Jaroszewicz J, Dentini M, Barbetta A, Brinchmann JE, Święszkowski W. Biofabrication. 2016;8:035002.
117. Gao B, Yang Q, Zhao X, Jin G, Ma Y, Xu F. Trends Biotechnol. 2016;34:746.
118. Hsieh FY, Lin HH, Hsu SH. Biomaterials. 2015;71:48.
119. Shim J-H, Jang K-M, Hahn SK, Park JY, Jung H, Oh K, Park KM, Yeom J, Park SH, Kim SW, Wang JH, Kim K, Cho D-W. Biofabrication. 2016;8:014102.
120. Huang S, Yao B, Xie J, Fu X. Acta Biomater. 2016;32:170.
121. Huang Y, Zhang XF, Gao G, Yonezawa T, Cui X. Biotechnol J. 2017;12:1600734. https://doi.org/10.1002/biot.201600734.

122. Martin I, Wendt D, Heberer M. Trends Biotechnol. 2004;22:80.
123. Chen HC, Hu YC. Biotechnol Lett. 2006;28:1415.
124. Stevens MM, Marini RP, Schaefer D, Aronson J, Langer R, Shastri VP. Proc Natl Acad Sci U S A. 2005;102:11450.
125. Kim J, Ma T. Tissue Eng Part A. 2012;18:120719071626004.
126. Heher P, Maleiner B, Prüller J, Teuschl AH, Kollmitzer J, Monforte X, Wolbank S, Redl H, Rünzler D, Fuchs C. Acta Biomater. 2015;24:251.
127. Moreira R, Velz T, Alves N, Gesche VN, Malischewski A, Schmitz-Rode T, Frese J, Jockenhoevel S, Mela P. Tissue Eng Part C Methods. 2015;21:530.
128. Radisic M, Marsano A, Maidhof R, Wang Y, Vunjak-Novakovic G. Nat Protoc. 2008;3:719.
129. Lee V, Singh G, Trasatti J, Bjornsson C, Xu X, Tran T, Yoo S, Dai G, Karande P. Tissue Eng Part C. 2014;20:473.
130. Jeon H, Kang K, Park SA, Kim WD, Paik SS, Lee S-H, Jeong J, Choi D. Gut Liver. 2017;11:121.
131. Miri AK, Mostafavi E, Khorsandi D, Hu SK, Malpica M, Khademhosseini A. Biofabrication. 2019;11:042002. https://doi.org/10.1088/1758-5090/ab2798.
132. Schöneberg J, De Lorenzi F, Theek B, Blaeser A, Rommel D, Kuehne AJC, Kießling F, Fischer H. Sci Rep. 2018;8:1.
133. Campos DFD, Lindsay CD, Roth JG, Lesavage BL, Seymour AJ, Krajina BA, Ribeiro R, Costa PF, Blaeser A, Heilshorn SC. Front Bioeng Biotechnol. 2020;8:1.
134. Zhang YS, Arneri A, Bersini S, Shin S-R, Zhu K, Goli-Malekabadi Z, Aleman J, Colosi C, Busignani F, Dell'Erba V, Bishop C, Shupe T, Demarchi D, Moretti M, Rasponi M, Dokmeci MR, Atala A, Khademhosseini A. Biomaterials. 2016;110:45.
135. Homan KA, Kolesky DB, Skylar-Scott MA, Herrmann J, Obuobi H, Moisan A, Lewis JA. Sci Rep. 2016;6:34845.
136. Horváth L, Umehara Y, Jud C, Blank F, Petri-Fink A, Rothen-Rutishauser B. Sci Rep. 2015;5:7974.
137. Dai X, Ma C, Lan Q, Xu T. Biofabrication. 2016;8:045005.
138. Zhou X, Zhu W, Nowicki M, Miao S, Cui H, Holmes B, Glazer RI, Zhang LG. ACS Appl Mater Interfaces. 2016;8:30017.
139. Zhang B, Montgomery M, Chamberlain MD, Ogawa S, Korolj A, Pahnke A, Wells LA, Masse S, Kim J, Reis L, Momen A, Nunes SS, Wheeler AR, Nanthakumar K, Keller G, Sefton MV, Radisic M. Nat Mater. 2016;15:669.
140. Lind JU, Busbee TA, Valentine AD, Pasqualini FS, Yuan H, Yadid M, Park S, Kotikian A, Nesmith AP, Campbell PH, Vlassak JJ, Lewis JA, Parker KK. Nat Mater. 2017;16:303. https://doi.org/10.1038/NMAT4782.
141. Skardal A, Devarasetty M, Kang H-W, Mead I, Bishop C, Shupe T, Lee SJ, Jackson J, Yoo J, Soker S, Atala A. Acta Biomater. 2015;25:24.
142. Ma X, Qu X, Zhu W, Li Y, Yuan S, Zhang H, Liu J, Wang P. Proc Natl Acad Sci. 2016;113:2206.
143. Lee H, Cho D-W. Lab Chip. 2016;16:2618.
144. Nguyen DG, Funk J, Robbins JB, Crogan-Grundy C, Presnell SC, Singer T, Roth AB. PLoS One. 2016;11:1.
145. Gu Q, Tomaskovic-Crook E, Lozano R, Chen Y, Kapsa RM, Zhou Q, Wallace GG, Crook JM. Adv Healthc Mater. 2016;5:1429.

146. Hou X, Liu S, Wang M, Wiraja C, Huang W, Chan P, Tan T, Xu C. SLAS Technol Transl Life Sci Innov. 2017;22:447. https://doi.org/10.1177/2211068216655753.
147. Zhang YS, Davoudi F, Walch PDK, Manbachi A, Luo X, Dell'Erba V, Miri A, Albadawi H, Arneri A, Li X, Wang X, Dokmeci M, Khademhosseini A, Oklu R. Lab Chip. 2016;16:4097.
148. Paul SM, Mytelka DS, Dunwiddie CT, Persinger CC, Munos BH, Lindborg SR, Schacht AL. Nat Rev Drug Discov. 2010;9:203.
149. Zhang YS, Arneri A, Bersini S, Shin S-R, Zhu K, Malekabadi ZG, Aleman J, Colosi C, Busignani F, Dell'Erba V. Biomaterials. 2016;110:45.

Non-Destructive and Label-Free Monitoring of 3D Cell Constructs

10

Hesham K. Yosef and Karin Schütze

Contents

What You Will Learn in This Chapter

Monitoring cells in 3D culture is a big challenge in cell biology and biomedical research. In this chapter, we will demonstrate the great potential of Raman trapping microscopy coupled with multivariate statistical analysis, to monitor individual cells

(continued)

H. K. Yosef · K. Schütze (✉)
CellTool GmbH, Lindemannstraße, Tutzing, Germany
e-mail: k.schuetze@celltool.de

© Springer Nature Switzerland AG 2021
C. Kasper et al. (eds.), *Basic Concepts on 3D Cell Culture*, Learning Materials in Biosciences,
https://doi.org/10.1007/978-3-030-66749-8_10

in 2D cultures and 3D tissues. We shortly describe the concept of spontaneous Raman spectroscopy using a bio-compatible 785 nm laser and the advantages of combined Optical Trapping. Furthermore, you will learn how to extract the rich chemical information from the Raman spectra using a unique Raman data analyzing software dedicated to biomedical needs. We show that Raman allows to monitor composition, functionality and quality of keratinocytes and fibroblasts during production as well as within the final graft. Furthermore, Raman allows to observe cell vitality within microspheres of mouse embryonic stem cells giving the potential to monitor drug penetration.

10.1 Raman Microscopy in Cell Biology

There is a great demand in cell biology and biomedical research for a non-invasive, non-destructive, and highly sensitive method for cell recognition. An ideal method can analyse and monitor biochemical changes in living cells under normal physiological conditions, without inducing cellular stress. Such vital properties are not met in the current analytical standards such as flow cytometry, immunofluorescence, and mass spectrometry [1]. It is well established that the application of antibodies, fluorescent stains, and the exposure to fluorescence excitation light damages live cells and induces phototoxicity [2, 3]. In contrast, Raman microscopy is a non-invasive and label-free optical method that provides a photonic fingerprint based on the chemical structural information of the analysed biomaterial, without the need of any contrasting agent such as fluorescence staining. Moreover, Raman excitation laser of 785-nm wavelength does not induce any cellular stress or toxicity, allowing a non-destructive analysis of live cells in culture [4, 5]. In addition, Raman measurements of a few cell-number (<100) provide meaningful results of the sample. Therefore, Raman microscopy has emerged as a promising tool for the analysis of cells in 2D cultures or 3D tissues [6, 7].

Raman is an inelastic light scattering phenomenon that arises upon the interaction of photons with chemical bondings of the targeted molecules. This interaction leads to molecular polarization, in which the charge distribution of the molecular bonds are changing, resulting in stimulated vibrations with subsequent scattering of the photons, as depicted in Fig. 10.1. The scattered Raman photons have a different energy than the incident photons, shifting to higher (Stokes scattering) or lower (anti-Stokes scattering) wavelength [8]. In spontaneous Raman, the Stokes scattered photons are collected and translated to Raman spectra, which can provide rich chemical information about the cellular content.

The Raman spectra consist of Raman bands and each band represents a specific vibrational signature of the cellular macromolecules such as protein, lipids, and DNA. A single molecule can have many signature bands along with the Raman spectra. For

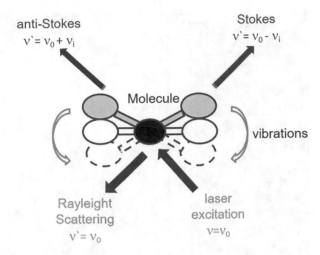

Fig. 10.1 Cartoon representation of the laser excitation of molecular bonds, inducing vibrations. This effect is accompanied by elastic scattering (Raylight) and inelastic Raman scattering (Stokes and anti-Stokes) of photons

instance, cellular proteins can give the following Raman bands; 756 (proteins: symmetric breathing of tryptophan), 1002 (proteins: C—C aromatic ring stretching of phenylalanine), 1044 (protein: proline), 1254 (protein: C—N in-plane stretching), 1345 (proteins: CH_3CH_2 wagging mode), 1453 (protein: CH_2CH_3 deformation), 1618 (proteins: C=C stretching mode of phenylalanine, tyrosine, tryptophan), 1657 cm^{-1} (protein: Amide C=O stretching for α-helix) as depicted in Fig. 10.2 [9, 10].

Take Home Message

Raman spectroscopy gives new insight into cell behaviour and development. In contrast to currently applied cell analysis methods, Raman microscopy is highly sensitive and specific, working in a label-free, fast, and sample sparing manner.

10.2   Raman Microscopic Setup

In this study, we implemented an inverted Raman microscope (BioRam®, CellTool GmbH, Tutzing, Germany) with 785 nm excitation laser. The inverted microscopic setup optimizes biological analysis since it allows applying different shapes and sizes of sample holders and providing more space to apply an incubator on top of the microscopic stage for environmental control. It also eliminates the risk of contamination, avoiding immersion of the objective lens into the cell culture media as required in an upright setup. In the BioRam® inverted setup, sample holders such as culture dishes, multi-well slides, or channel-slides with coverglass bottom are used. Due to especial laser coupling, BioRam® possesses integrated trapping features that arrest cells within the laser focus during Raman analysis [11]. In addition, the spectral intensity is multiplied that enables the acquisition of high-quality Raman spectra even from the depth of tissue samples, depicting a 3D biochemical

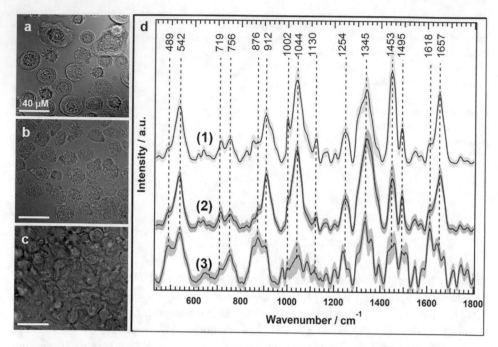

Fig. 10.2 Bright field micrograph of keratinocytes (**a**), fibroblasts (**b**), and melanocytes (**c**). The Raman mean spectra (panel d) of keratinocytes (1), fibroblasts (2), and melanocytes (3) are collected from 300 keratinocytes (3 donors), 300 fibroblasts (3 donors), and 100 melanocytes cells (1 donor). Raman measurements were collected using 785 nm laser of the Raman Trapping Microscope BioRam® (CellTool GmbH, Tutzing, Germany). The spectra were smoothed, baseline corrected, and unit-vector normalized. This figure is adopted from H K.Yosef et al. [10] (Permission is obtained from springer nature)

distribution. Moreover, optical trapping allows the measurement of non-adherent cells within their physiological environment, eliminating the need to attach them before analysis i.e. using cytospin techniques or drying them onto the slide.

Take Home Message

The integration of the two, Nobel prize awarded highly complex photonic technologies—Raman spectroscopy and Optical Trapping—into an inverted digital microscope makes this revolutionary technology available for biomedical research and application. Cells are analysed within their native environment reflecting their natural features and behaviour, they can be measured repeatedly, and kept available for downstream applications.

10.3   Confocal Raman Microscopy

The principles of confocal microscopy were first described and developed by Marvin Minsky (1961), which is now the basis of all modern confocal microscopes [12]. Since then confocal microscopes have become a popular tool for visualizing biological specimens. It has increased the resolution of laser and fluorescence microscopy. The concept of confocal measurements is depending on focusing a point-shaped light through an objective into the sample. The projected spot is then focused back by the same objective, through a small aperture called pinhole onto a highly sensitive detector. Only photons that arise from the focal plane can pass through the pinhole as depicted in Fig. 10.3 [12, 13]. Collection of Raman photons using the confocal setup significantly enhances the measurement resolution and selectivity. Due to that confocal setup BioRam® allows the collection of photons from a small volume (confocal volume) of the cell without interference with the surrounding parts of the same cell, such as measuring cell nucleus or cytoplasm. Even more, it enables to selectively measure smaller intracellular structures such as cytoplasmic

Fig. 10.3 Cartoon representation of laser focusing on a biological cell and acquisition of the scattered Raman photons in a confocal setup. The pinhole allows only the scattered photons that arise from the focus point to reach the detector

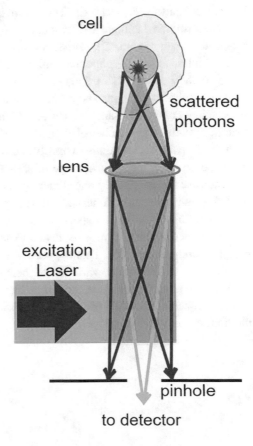

vacuoles, lipid droplets, or nucleolus. Furthermore, the confocal setup decreases the fluorescence background that can overlap with the Raman scatterings, by limiting the detection to photons derived only from the confocal volume [13, 14]. Moreover, the confocal Raman measurements of different layers in the z-direction (measurements at different vertical layers) can be integrated to acquire a 3D map of the biochemical distribution in the 3D biological model.

> **Take Home Message**
> Applying confocal setup increases the selectivity and resolution of Raman measurements, by limiting the collection of Raman photons to the confocal volume of the laser focus.

10.4   Multivariate Statistical Analysis

The system is also coupled with a data analysis platform CT-RamSES (CellTool Raman Statistical Analysis Software, CellTool GmbH, Tutzing, Germany). In a push-of-a-button, this platform can facilitate the extraction of biomedical relevant information from the acquired Raman spectra. CT-RamSES can run all required statistical procedures to compare samples, to detect subpopulation, and to identify the biochemical cellular changes based on the Raman spectra. This is achieved by the automatic processing of the spectra such as baseline corrections and normalization steps, as well as statistical evaluations such as principal components analysis (PCA) and hierarchical cluster analysis (HCA). Raman spectra provide large number of digital data points. With BioRam® standard acquisition, around 973 pairs of data points (intensity and wavenumber at each point) are collected for a single spectrum at the spectral range of 350–3300 cm^{-1}. Collecting thousands of spectra will increase the complexity and dimensions of the data. Therefore, multivariate analysis methods are implemented to analyse these large Raman data sets, illuminating differences, and similarities amongst the spectral data. Principal components analysis (PCA) can be used to reduce the large number of data set variables to smaller number of variables called principal components [15]. This approach can display the variation between the samples and cell types based on the variation of their Raman spectra, which can be shown in the PCA score plot. Furthermore, applying hierarchical cluster analysis (HCA) allows to separate subclasses of data of one sample, based on similarity of the spectra, to evaluate the homogeneity within cell population.

Take Home Message

Raman spectra are composed of several thousands of data information. Statistical evaluation is required to depict cell features or change in behaviour. CT-RamSES was developed from biologists to allow even the statistical unexperienced to get easy access to biological relevant data analysis and providing biomedical meaningful interpretation just on the-push-of-a-button.

10.5   Raman Monitoring of 2D and 3D Cell Cultures

In this chapter, the potential of Raman microscopy to monitor cells in the 3D cultures is presented. Two major types of 3D cultures will be addressed: (1) Cultures constructed on artificial scaffold such as skin cells on a hydrogel matrix to form a skin graft. (2) Cultures constructed without scaffold such as microspheres assembled in special microplates.

The autologous skin graft is used as a replacement of injured skin caused by trauma or surgical interventions. The construction of the graft requires the expansion of fibroblasts and keratinocytes that are isolated from the patient and cultured separately in normal cell cultures (2D). Currently, FACS (Fluorescence-activated cell sorting) is used at this step as quality control to monitor cross-contamination between the separately cultured keratinocytes and fibroblasts. However, it is an invasive technique as cells change their features upon antibody labelling and it requires large number of cells (approx. 10^6), which cannot be used for patient treatment. Moreover, there is a lack of specific antibodies for fibroblasts discrimination [5]. Next, fibroblasts are mingled with hydrogel and allowed to grow and expand for 5 days. Then, keratinocytes are poured on top of the 3D hydrogel matrix forming a stable single cell layer after another 5 days of growth, as depicted in Fig. 10.4. Currently, quality control of the final product is performed using DNA count of a 1×1 cm piece. However, DNA count provides only a total number of cells, without discriminating cell types, cross-contamination, or providing information about cell vitality/functionality [16–18]. Thus, BioRam® provides improved quality control measures, ensuring minimal cross-contamination whilst shortening valuable time for patient treatment.

Take Home Message

Raman microscopy is superior to standard quality control methods allowing to check cross-contamination and monitor cell condition in both, 2D and 3D cultures, ensuring quality of the graft whilst shortening valuable time for patient treatment, minimizing production costs and securing patient safety.

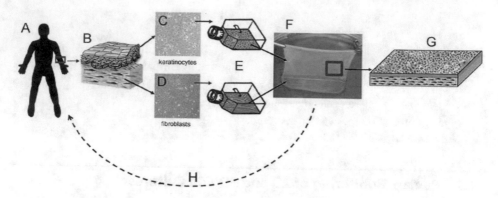

Fig. 10.4 Schematic description of skin graft production and treatment steps. Skin sample is collected from patient (**a**), cells are extracted from different layers of the patient skin (**b**) to separate cultures of keratinocytes (**c**) and fibroblasts (**D**). These cells are expanded in the separate cultures (**e**) then the fibroblasts are seeded on the hydrogel and cultured for 5 days, followed by adding keratinocytes on the top of the fibroblasts (**F**) and incubate them for another 5 days to form the final skin graft product (**G**), which can be transplanted to the patient (**H**). The figure is courtesy of Prof. Ernst Reichmann (Tissue Biology Research Unit, University Children's Hospital, Zurich, Switzerland)

10.6   Raman Trapping Microscopy to Monitor Cells in 2D Culture

Raman measurements were conducted on 2D pure cultures of keratinocytes, fibroblasts, and melanocytes. Individual skin cells were trapped and kept within the Raman laser focus during Raman spectra acquisition. The collected Raman spectra reveal characteristic differences between the spectral patterns for each cell type as depicted in Fig. 10.2. The Raman bands around 489 (glycogen), 719 (DNA/RNA: adenine; lipid), 876 (collagen; lipid), 1002 (protein: phenylalanine), 1044 (collagen: proline), 1453 (protein; lipids), 1618 (proteins: phenylalanine, tyrosine, tryptophan), and 1657 cm^{-1} (protein) are showing the most drastic differences among the three cell types [9, 10]. These differences in the spectral patterns are due to different compositions of biomolecules among cell types, such as the differences in cellular expressions of proteins [19, 20].

Take Home Message
These results indicate the advantage of Raman microscopy to identify skin cells in 2D cell cultures providing characteristic pattern for each cell typ. This way the percentage of cross-contamination amongst the fibroblasts and keratinocytes can be evaluated in a fast and reliable manner.

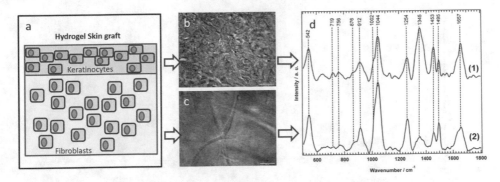

Fig. 10.5 Raman measurements of cells in the skin graft matrix. Cartoon representation of the skin graft structure (**a**), Bright field photos of keratinocytes layer of the graft (**b**), Bright field photos of fibroblasts layer of the graft (**c**), and mean Raman spectra (**d**) of keratinocytes (d1) and fibroblasts (d2)

10.7   Raman Microscopy to Monitor Cells in 3D Grafts

After expanding the cells in 2D culture, fibroblasts are mixed with hydrogel, poured on an 8×8 cm^2 shape, and allowed to grow and expand. Keratinocytes are seeded on top of the hydrogel-fibroblast matrix creating two separate layers of the matrix as depicted in Fig. 10.5a. The graft is cultured for 11 days to reach high cellular growth and attachment to the matrix.

10.8   Purity of Expanded Cells

Raman measurements were collected from keratinocytes at the top layer of the graft and from fibroblasts within the 3D-matrix. As revealed in Fig. 10.5, the Raman spectra of both keratinocytes and fibroblasts are showing a characteristic pattern of each type. Raman pattern of fibroblast within 3D-graft (Fig. 10.4 d2) differ from fibroblast in 2D culture (Fig. 10.3 d2), which seems to be related to their pronounced difference in shape and also the effect cell-matrix contact (Fig. 10.3b and Fig. 10.4c). Raman pattern of fibroblast within 3D-graft differs significantly from keratinocytes as depicted in Fig. 10.5d and Fig. 10.2d, respectively. The most pronounced differences between keratinocytes and fibroblasts are the Raman bands at 1345 and 1453 cm^{-1}. These two bands have higher intensities in case of keratinocytes, indicating differences in the amount of lipid and protein content between keratinocytes and fibroblasts. This spectral feature was observed in both culture types: 2D and 3D cultures (Fig. 10.2d, and Fig. 10.5d). To clearly visualize the Raman spectral variations, PCA was conducted on Raman spectra collected from the graft. The PCA-scores plot reveals different scattering patterns of the spectra collected from the two cell types as depicted in Fig. 10.6. Therefore, Raman spectra can be used to detect, monitor, and classify keratinocytes and fibroblasts in the 3D skin graft matrix. Raman spectra can be collected at

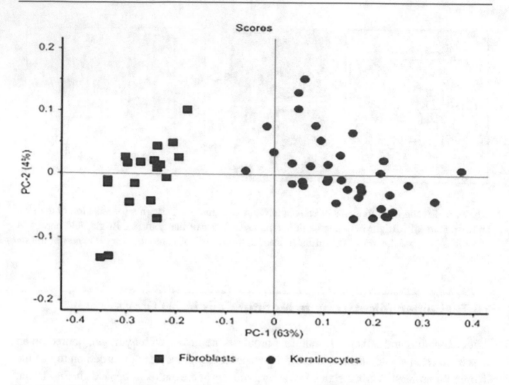

Fig. 10.6 PCA score plot (PC1 VS PC2) of keratinocytes (red dots) and fibroblasts (blue dots) based on the Raman data of Fig. 5d, indicating clear differences between the two cell types

different depths (0, 40, 130, 170, 200, 220, and 260μm) of the fibroblast layers of the skin graft, without great loss in spectral quality, as depicted in Fig. 10.7.

10.9   Purity of Skin Graft

To test the potential of this approach, Raman measurements were conducted to evaluate the purity of the graft. The graft was placed up-side-down in which the keratinocyte layers are in direct contact with a borosilicate cover glass slide (the cover glass slide is used as a substrate for the Raman analysis). Raman spectra were then collected from the keratinocyte layer of the graft. As displayed in Fig. 10.8a, 3861 Raman spectra were collected from 561 cells in the keratinocytes layer. Cluster analysis was applied on this data set using HCA algorithm. HCA classified three basic clusters: Cluster 1 (1356 spectra), cluster 2 (2176 spectra), and cluster 3 (329 spectra), as shown in Fig. 10.8b. Clusters 1 and 2 (Fig. 10.8, Panels 1, 2) are similar to keratinocytes, their Raman spectra exhibit minor differences in bands intensity that could be related to different cellular stages of the keratinocytes. In contrast, cluster 3 (Fig. 10.8, Panel 3) is showing more distinctive differences and the Raman band at 1453 cm^{-1} is showing lower intensity compared to the other 2 clusters. By

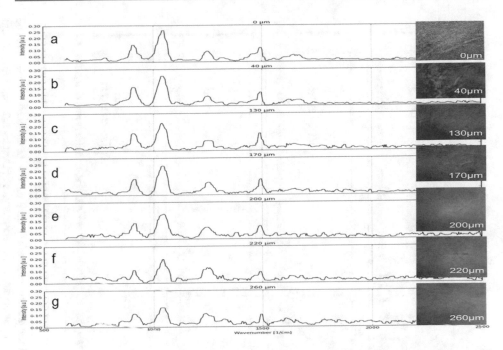

Fig. 10.7 Bright field photos and respective Raman spectra collected at: (**a**) surface—0μm, (**b**) 40μm, (**c**) 130μm, (**d**) 170μm, (**e**) 200μm, (**f**) 220μm, and (**g**) 260μm depth of the skin graft

comparing the spectra of cluster 3 with the spectra of fibroblasts, a clear similarity was observed (Fig. 10.8, Panels 3, 4). Therefore, Raman spectral results indicate around 8.5% contamination of fibroblasts in the keratinocyte layer. Moreover, Raman can discriminate healthy from apoptotic cells. We compared fresh skin grafts with one-week-old samples. The Raman bands intensities of proteins and nucleic acids of live cells were clearly deteriorated in case of the old sample, which can be attributed to fibroblast cell death and the graft decay (Fig. 10.9) [21].

Take Home Message
These results clearly indicate the strong potential of Raman microscopy followed by especial statistical evaluation methods to become a standard in quality control of cell-based therapeutics, monitoring cellular condition and identifying cross-contamination during production and of the final graft, in an efficient and sampling sparing manner.

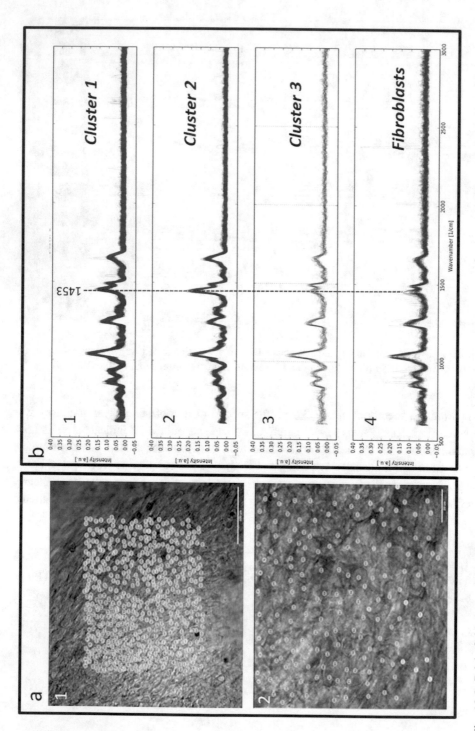

Fig. 10.8 Bright field photo of the keratinocytes layer (a1) and zooming of the measurement region (a2) using 60× water immersion objective. The green marks are indicating Raman measurement points. Results of HCA clustering of Raman data from keratinocytes of clusters 1, 2, and 3 (b1, b2, and b3). Raman measurements of fibroblasts (b4)

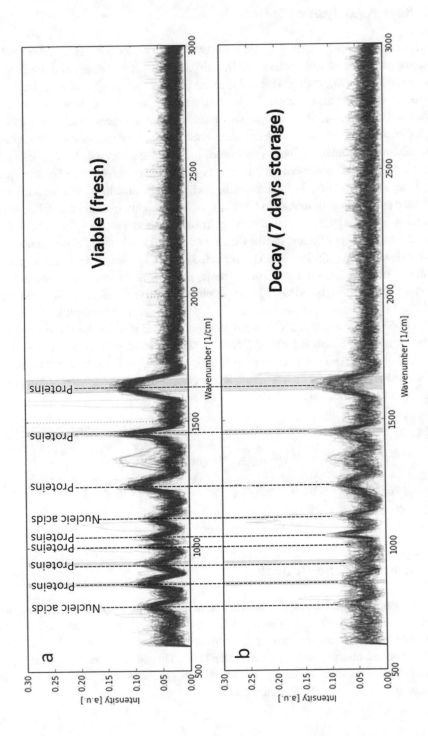

Fig. 10.9 Raman spectra of fibroblasts within the graft (**a**) freshly prepared and (**b**) after 7 days of storage at room temperature

10.10 Raman Analysis of Spheroids

Spheroids have emerged recently as an attractive 3D cell-model to recapitulate tissue biology, complexity, and architecture in 3D, unlike 2D in vitro systems, and are gaining increasing interest in drug testing. This is due to the fact that cells in spheroids are behaving like in tissue environment since the extracellular matrix has a great influence on cells behaviour [22]. Therefore, Raman analysis of such an environment would be useful to collect information of the cells at the circumference and in the depth of the microsphere. Raman microscopic analysis was conducted on microspheres of 150µm diameter constructed from mouse embryonic stem cells (Fig. 10.10a), cultured in the Sphericalplate 5D for 3 days (Kugelmeiers Ltd., Erlenbach, Switzerland), enabling the formation of a regular-sized spheroid (as described in Chap. 3), which is highly instrumental for developing a scaling-up strategy in optical screening in order to analyse the spheroids. Raman spectra were collected at different depths (0, 25, and 50µm) without any significant loss of Raman bands intensity (Fig. 10.10b). The collected Raman spectra reveal the biochemical compositions of cells at each layer. For example, *proteins features* are detected around: 622 (phenylalanine), 641 (tyrosine), 759 (tryptophan), 856 (tyrosine/collagen), 1005 (phenylalanine), 1175 (cytosine), 1246 (amide III), 1449 (CH deformation/lipid), 1609 (cytosine), and 1662 cm^{-1} (amide I/DNA). *DNA features* are indicated by the Raman bands at 726 (adenine), 787 (phosphodiester of DNA backbone), 1096 (phosphodioxy group of DNA), 1340 (nucleic acid mode), 1578 (guanine), and 1662 cm^{-1} (nucleic acid modes/ proteins). Furthermore, a very characteristic Raman *feature of carbohydrates* is detected at 942 cm^{-1} (polysaccharides skeletal mode).

Take Home Message
These results indicate that Raman microscopy is able to detect the biochemical profile of cells even in the depth of microspheres, providing the opportunity to trace and evaluate changes of cells in response to environmental impact or to compound treatment.

10.11 Conclusion and Outlook

The unique features of the BioRam® platform (i.e. inverted microscopy, laser trapping, and complete data analysis workflow) grant biologists and medical personnel easy access to the advantages of Raman micro-spectroscopy, enabling label-free and non-invasive cell analysis for biomedical routines and research applications. This allows deep insight into cells from a multi-omics perspective gaining useful information about behaviour, development, or fate of cells in a fast and efficient manner.

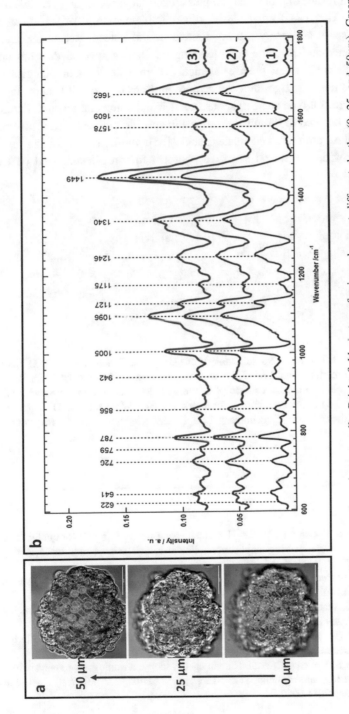

Fig. 10.10 Raman analysis of mouse embryonic stem cells, Bright field photos of microspheres at different depths (0, 25, and 50μm). Green marks indicate Raman measurement points, the scale bar is covering 40μm. Raman mean spectra of the cells measured within the microsphere at 0, 25, and 50μm depths (b1, b2, and b3)

Raman spectral results of biological samples are digital data carrying valuable biochemical information. The data format allows direct implementation of statistical analysis algorithms to classify and analyse these Raman digital values. Recent applications of artificial intelligence and machine deep learning on Raman spectral data has opened a new era of fast data analysis and precise chemical pattern recognition of biological samples [23, 24]. Machine learning algorithm can acquire millions of spectral features that can be used to characterize cell type and cell status in a fast and automatic manner. Furthermore, this approach can be used to illuminate changes in specific molecular contents within the cells such as lipids, proteins, and nucleic acids [25]. Furthermore, Raman analysis of spheroids can provide a fast and sensitive platform for drug testing and personalized medicine.

Take Home Message
The combination of Raman microscopy and digital data analysis will tremendously speed up and enhance quality control of cell therapeutics and revolutionize diagnosis of disease assisting the physicians in treatment decision. In addition, sensitive monitoring of drug response on tumour spheroids will support drug development—both to improve healthiness and save life.

Acknowledgement We would like to thank Prof. Dr. Ernst Reichmann, TBRU (University-Children's Hospital Zürich, Switzerland) as well as Dr. Daniela Marino CUTISS AG (Zürich, Switzerland) for providing the skin cell samples. We would like also to thank Sarvesh Ghorpade, Dr. Heidi Kremling and Florian Zunhammer for their assistance. We thank also Dr. Ali Mirsaidi and Dr. Patrick Kugelmeier from Kugelmeiers Ltd. (Erlenbach, Switzerland) for providing the microspheres and for their great support.

References

1. Pudlas M, Koch S, Bolwien C, Walles H. Raman spectroscopy as a tool for quality and sterility analysis for tissue engineering applications like cartilage transplants. Int J Artif Organs. 2010;33:228–37.
2. Jensen EC. Use of fluorescent probes: their effect on cell biology and limitations. Anat Rec Adv Integr Anat Evol Biol. 2012;295:2031–6. https://doi.org/10.1002/ar.22602.
3. Icha J, Weber M, Waters JC, Norden C. Phototoxicity in live fluorescence microscopy, and how to avoid it. BioEssays. 2017;39:1700003. https://doi.org/10.1002/bies.201700003.
4. Notingher I, Verrier S, Romanska H, et al. In situ characterisation of living cells by Raman spectroscopy. Spectroscopy. 2002;16:43–51. https://doi.org/10.1155/2002/408381.
5. Pudlas M, Koch S, Bolwien C, et al. Raman spectroscopy: a noninvasive analysis tool for the discrimination of human skin cells. Tissue Eng Part C Methods. 2011;17:1027–40. https://doi.org/10.1089/ten.tec.2011.0082.

6. Charwat V, Schütze K, Holnthoner W, et al. Potential and limitations of microscopy and Raman spectroscopy for live-cell analysis of 3D cell cultures. J Biotechnol. 2015;205:70–81. https://doi.org/10.1016/j.jbiotec.2015.02.007.

7. Steinke M, Gross R, Walles H, et al. An engineered 3D human airway mucosa model based on an SIS scaffold. Biomaterials. 2014;35:7355–62. https://doi.org/10.1016/j.biomaterials.2014.05.031.

8. Raman CV, Krishnan KS. A new type of secondary radiation. Nature. 1928;121:501–2. https://doi.org/10.1038/121501c0.

9. Talari ACS, Movasaghi Z, Rehman S, ur RI. Raman spectroscopy of biological tissues. Appl Spectrosc Rev. 2015;50:46–111. https://doi.org/10.1080/05704928.2014.923902.

10. Pörtner R, editor. Animal cell biotechnology: methods and protocols, methods in molecular biology. Cham: Springer; 2020.

11. Ashkin A, Dziedzic JM, Yamane T. Optical trapping and manipulation of single cells using infrared laser beams. Nature. 1987;330:769–71. https://doi.org/10.1038/330769a0.

12. Minsky M. Memoir on inventing the confocal scanning microscope: memoir on inventing the confocal scanning microscope. Scanning. 1988;10:128–38. https://doi.org/10.1002/sca.4950100403.

13. Smith R, Wright KL, Ashton L. Raman spectroscopy: an evolving technique for live cell studies. Analyst. 2016;141:3590–600. https://doi.org/10.1039/C6AN00152A.

14. Conchello J-A, Lichtman JW. Optical sectioning microscopy. Nat Methods. 2005;2:920–31. https://doi.org/10.1038/nmeth815.

15. Bonnier F, Byrne HJ. Understanding the molecular information contained in principal component analysis of vibrational spectra of biological systems. Analyst. 2012;137:322–32. https://doi.org/10.1039/C1AN15821J.

16. Larouche D, Cantin-Warren L, Desgagné M, et al. Improved methods to produce tissue-engineered skin substitutes suitable for the permanent closure of full-thickness skin injuries. BioResearch Open Access. 2016;5:320–9. https://doi.org/10.1089/biores.2016.0036.

17. MacNeil S. Progress and opportunities for tissue-engineered skin. Nature. 2007;445:874–80. https://doi.org/10.1038/nature05664.

18. Marino D, Luginbuhl J, Scola S, et al. Bioengineering dermo-epidermal skin grafts with blood and lymphatic capillaries. Sci Transl Med. 2014;6:221ra14. https://doi.org/10.1126/scitranslmed.3006894.

19. Per Oksvold MG. Tissue-specific protein expression in human cells, tissues and organs. J Proteomics Bioinform. 2010;03(10):294–301. https://doi.org/10.4172/jpb.1000153.

20. Pontén F, Gry M, Fagerberg L, et al. A global view of protein expression in human cells, tissues, and organs. Mol Syst Biol. 2009;5:337. https://doi.org/10.1038/msb.2009.93.

21. Brauchle E, Thude S, Brucker SY, Schenke-Layland K. Cell death stages in single apoptotic and necrotic cells monitored by Raman microspectroscopy. Sci Rep. 2015;4:4698. https://doi.org/10.1038/srep04698.

22. Langhans SA. Three-dimensional in vitro cell culture models in drug discovery and drug repositioning. Front Pharmacol. 2018;9:6. https://doi.org/10.3389/fphar.2018.00006.

23. Ho C-S, Jean N, Hogan CA, et al. Rapid identification of pathogenic bacteria using Raman spectroscopy and deep learning. Nat Commun. 2019;10:4927. https://doi.org/10.1038/s41467-019-12898-9.

24. Krauß SD, Roy R, Yosef HK, et al. Hierarchical deep convolutional neural networks combine spectral and spatial information for highly accurate Raman-microscopy-based cytopathology. J Biophotonics. 2018;11:e201800022. https://doi.org/10.1002/jbio.201800022.

25. Talari ACS, Rehman S, Rehman IU. Advancing cancer diagnostics with artificial intelligence and spectroscopy: identifying chemical changes associated with breast cancer. Expert Rev Mol Diagn. 2019;19:929–40. https://doi.org/10.1080/14737159.2019.1659727.

Further Reading

Chapter (Raman Trapping Microscopy for Non-invasive Analysis of Biological Samples). Animal cell biotechnology: methods and protocols, methods in molecular biology. Cham: Springer; 2020.

Steinke M, Gross R, Walles H, et al. An engineered 3D human airway mucosa model based on an SIS scaffold. Biomaterials. 2014;35:7355–62.

Charwat V, Schütze K, Holnthoner W, et al. Potential and limitations of microscopy and Raman spectroscopy for live-cell analysis of 3D cell cultures. J Biotechnol. 2015;205:70–81.

Pudlas M, Koch S, Bolwien C, Walles H. Raman spectroscopy as a tool for quality and sterility analysis for tissue engineering applications like cartilage transplants. Int J Artif Organs. 2010;33:228–37.

Index

Printed in the United States
by Baker & Taylor Publisher Services